SPRINGER TRACTS IN MODERN PHYSICS

Ergebnisse der exakten Naturwissenschaften

Volume 42

Editor: G. Höhler

Springer-Verlag Berlin Heidelberg GmbH 1966

Manuscripts for publication should be addressed to:

G. HÖHLER, Institut für Theoretische Kernphysik der Technischen Hochschule, 75 Karlsruhe, Kaiserstraße 12

Proofs and all correspondence concerning papers in the process of publication should be addressed to:

E. A. NIEKISCH, Kernforschungsanlage Jülich, Arbeitsgruppe Institut für Technische Physik, 517 Jülich, Postfach 365

ISBN 978-3-662-15910-1 ISBN 978-3-540-34900-6 (eBook)
DOI 10.1007/978-3-540-34900-6

Originally published by Springer-Verlag, Berlin · Heidelberg in 1966
Softcover reprint of the hardcover 1st edition 1966

Library of Congress Catalog Card Number 25-91 30

Quadrupole Optics

(The electron optical properties of rectilinear orthogonal systems)

P. W. Hawkes

Contents

1. Introductory 1
2. The paraxial properties of orthogonal systems 1
3. Primary aberrations 13
   3.1 Variation of parameters 15
   3.2 Perturbation characteristics 19
   3.3 Integral equations 34
   3.4 Permissible aberrations and aberration patterns 42
   3.5 Mechanical aberrations 49
      a) "Das Auflösungsvermögen sphärisch korrigierter elektrostatischer Elektronenmikroskope" by *W. E. Meyer* 49
      b) The work of *J. H. M. Deltrap* on magnetic quadrupoles 53
      c) *A. V. Crewe's* study of the stigmatic magnetic doublet 55
      d) *P. F. Meads'* analysis of any failure to achieve quadrupole symmetry 56
      e) Comment 57
4. Values of the cardinal elements and aberration coefficients of quadrupole lenses 57
   4.1 Potential model 57
   4.2 Transfer matrices and cardinal elements 60
      a) The rectangular model 60
      b) The bell-shaped model 61
      c) The modified rectangular model 69
      d) The triangular model 70
      e) Practical measurements and calculations 72
   4.3 Aperture aberration coefficients 73
      a) The rectangular model 74
      b) The bell-shaped model 75
      c) Other models 79
      d) Measurements 79
   4.4 Lens systems 82
      a) The quadrupole doublet 82
      b) The quadrupole triplet 86
      c) Systems of four or more quadrupole lenses 87
5. Chromatic aberration 105
   5.1 Introductory 105
   5.2 Geometrical aberrations 106
   5.3 Disparities between the field distributions 117
      a) Small differences 117
      b) Large differences: separated lenses 118
6. Concluding remarks 119

References 120

1. Introductory

Round lenses have two drawbacks: for charged particles of high energy and for heavy particles, they become ineffectual, and their unavoidable spherical and chromatic aberration limits the resolution of instruments that rely upon them [155]. That quadrupoles offer a means of overcoming each of these difficulties was observed, quite independently, by *Christofilos* [30] and *Courant, Livingston* and *Snyder* [34] and by *Scherzer* [156]; quadrupoles had been studied formally, as the electron optical counterparts of toric lenses, even earlier, by *Melkich* [133]. The spheres of application and *raisons d'être* of the two types of quadrupoles – strong-focusing lenses for controlling high energy beams and quadrupole lenses for overcoming lens aberrations – are so very different that they have been studied largely independently. In the design of very high voltage electron microscopes, however, both potentialities of quadrupoles are likely to be exploited.

The terminology and attitude in this memoir are those appropriate to electron optics, but as the work of *Meads*, which will be described in detail, bears witness, the artificial division between electron optics and strong focusing is rapidly dwindling. Some attempt has thus been made to survey the electron optical literature very thoroughly; this is not true of strong focusing, on which several review articles and chapters have already been published [8b, 29, 83, 102, 115, 126, 138, 182, 197, 198]. The notation used is for the most part that of *Glaser* [74]; a note on German terminology has been prepared by *Scherzer* [159].

2. The paraxial properties of orthogonal systems

The paths of electrons in electric and magnetic fields may be deduced from Fermat's principle, which states that $\delta \int n\, \mathrm{d}s = 0$ when the integration is taken along a ray and the end-points are not varied; $\mathrm{d}s$ is an element of arc-length along the ray, and the refractive index, n, is given by

$$n = \sqrt{\varphi(1+\varepsilon\varphi)} - \eta\, \boldsymbol{A}\cdot\boldsymbol{s}\,.$$

$\varphi(x, y, z)$ is the scalar potential of the electric field, $\boldsymbol{E} = -\operatorname{grad}\varphi$, and $\boldsymbol{A}$ is the vector potential of the magnetic field, $\boldsymbol{B} = \operatorname{curl}\boldsymbol{A}$; $\boldsymbol{s}$ is a unit vector tangential to the trajectory, ε denotes the constant $e/2m_0c^2 \cong 10^{-6}$ C sec^2 kg^{-1} m^{-2}, put equal to zero in the non-relativistic approximation, and η denotes the constant $\sqrt{e/2m_0} = 2.965 \times 10^5$ C$^{1/2}$ kg$^{-1/2}$. The charge (in C) and rest mass (in kg) of the electron are denoted by $-e$ and m_0 respectively.

By a *rectilinear* system, we mean a system with a straight optic axis; if this is the z-axis of cartesian coordinate axes, we write $\int n\, \mathrm{d}s = \int m\, \mathrm{d}z$ and

$$m = \sqrt{\varphi(1+\varepsilon\varphi)(1+x'^2+y'^2)} - \eta(A_x x' + A_y y' + A_z) \qquad (2.1)$$

in which a prime indicates differentiation with respect to z. Writing $V_{\alpha\beta} = \int_{z_\alpha}^{z_\beta} m\,dz$ it is found that if the integration is taken along a ray,

$$\delta V_{\alpha\beta} = p_\beta\,\delta x_\beta + q_\beta\,\delta y_\beta - (p_\alpha\,\delta x_\alpha + q_\alpha\,\delta y_\alpha)$$

in which $p = \partial m/\partial x'$, $q = \partial m/\partial y'$. On performing the Legendre transformation

$$T_{\alpha\beta} = V_{\alpha\beta} + p_\alpha x_\alpha + q_\alpha y_\alpha - (p_\beta x_\beta + q_\beta y_\beta)\,,$$

we find that

$$\delta T_{\alpha\beta} = -(x_\beta\,\delta p_\beta + y_\beta\,\delta q_\beta) + x_\alpha\,\delta p_\alpha + y_\alpha\,\delta q_\alpha$$

so that $T_{\alpha\beta}$ is a function of p_α, q_α, p_β and q_β and, provided that p_β and q_β are not proportional to p_α and q_α respectively,

$$\begin{aligned} x_\beta &= -\frac{\partial T_{\alpha\beta}}{\partial p_\beta} \qquad & x_\alpha &= \frac{\partial T_{\alpha\beta}}{\partial p_\alpha}\,, \\ y_\beta &= -\frac{\partial T_{\alpha\beta}}{\partial q_\beta} \qquad & y_\alpha &= \frac{\partial T_{\alpha\beta}}{\partial q_\alpha}\,. \end{aligned} \tag{2.2}$$

The function $V(x_\alpha, y_\alpha, x_\beta, y_\beta)$ is Hamilton's point characteristic and $T(p_\alpha, q_\alpha, p_\beta, q_\beta)$ is his angle characteristic; we shall not require either of the mixed characteristics, which are obtained from V by two other Legendre transformations. The geometrical interpretation of T is shown in Fig. 1: it is the optical path-length between the feet of the perpendiculars from the points $(0, 0, z_\alpha)$ and $(0, 0, z_\beta)$ to the tangents to the rays at the points where they intersect the planes $z = z_\alpha$, $z = z_\beta$.

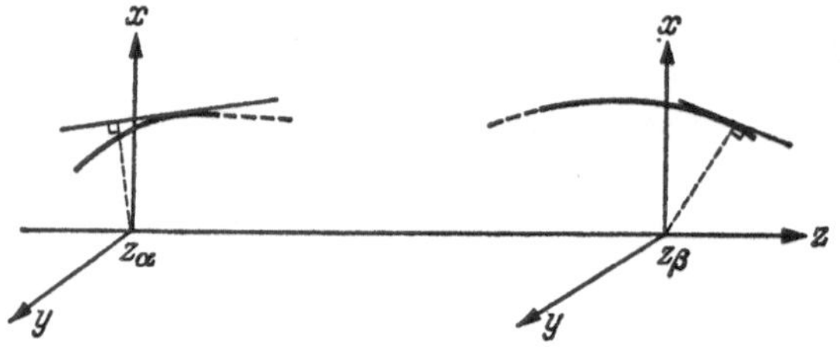

Fig. 1. The angle characteristic, T

The paraxial properties are obtained by writing φ and the components of $\boldsymbol{A}$ as power series in X and Y, and retaining only the terms of lowest degree in X, Y and their derivatives (XYz form an arbitrary cartesian triad). In a rectilinear system, only terms of even degree occur, and writing

$$\begin{aligned} \varphi(X, Y, z) = \Phi(z) &- \frac{1}{4}\{\Phi''(z) - D(z)\}\,X^2 + P(z)\,XY - \\ &- \frac{1}{4}\{\Phi''(z) + D(z)\}\,Y^2 + \left\{\frac{1}{64}\,\Phi^{(\mathrm{iv})}(z) - \frac{1}{48}\,D''(z) + D_1(z)\right\}X^4 - \\ &- \left\{\frac{1}{12}\,P''(z) - 4\,P_1(z)\right\}X^3\,Y + \left\{\frac{1}{32}\,\Phi^{(\mathrm{iv})}(z) - 6\,D_1(z)\right\}X^2\,Y^2 - \\ &- \left\{\frac{1}{12}\,P''(z) + 4\,P_1(z)\right\}X\,Y^3 + \left\{\frac{1}{64}\,\Phi^{(\mathrm{iv})}(z) + \frac{1}{48}\,D''(z) + D_1(z)\right\}Y^4 \end{aligned} \tag{2.3a}$$

$$A_X(X, Y, z) = -\Omega(z)\,Y + \frac{1}{4}\{\Omega''(z) + \Delta'(z)\}\,X^2 Y + \\ + \frac{1}{2}Q'(z)\,X Y^2 + \frac{1}{12}\{\Omega''(z) - \Delta'(z)\}\,Y^3$$

$$A_Y(X, Y, z) = 0$$

$$A_z(X, Y, z) = \frac{1}{2}Q(z)\,X^2 - \frac{1}{2}\{\Omega'(z) + \Delta(z)\}\,XY - \\ - \frac{1}{2}Q(z)\,Y^2 - \left\{\frac{1}{48}Q''(z) - Q_1(z)\right\}X^4 + \\ + \left\{\frac{1}{16}\Omega'''(z) + \frac{1}{12}\Delta''(z) - 4\Delta_1(z)\right\}X^3 Y + \\ + \left\{\frac{1}{8}Q''(z) - 6Q_1(z)\right\}X^2 Y^2 + \\ + \left\{\frac{1}{48}\Omega''' + 4\Delta_1(z)\right\}X Y^3 + \left\{\frac{1}{48}Q''(z) + Q_1(z)\right\}Y^4 \qquad (2.3b)$$

we obtain

$$m = m^{(0)} + m^{(2)} + m^{(4)} + \cdots$$

in which $m^{(r)}$ represents all the terms of degree r; $m^{(0)}$ and $m^{(2)}$ are given by the following formulae:

$$m^{(0)} = \sqrt{\Phi^*}$$

$$m^{(2)} = -\frac{1+2\varepsilon\Phi}{8\sqrt{\Phi^*}}\Phi''(X^2+Y^2) + \\ + \left(\frac{1+2\varepsilon\Phi}{8\sqrt{\Phi^*}}D - \frac{1}{2}\eta Q\right)(X^2 - Y^2) + \left(\frac{1+2\varepsilon\Phi}{2\sqrt{\Phi^*}}P + \frac{1}{2}\eta\Delta\right)XY + \\ + \frac{1}{2}\sqrt{\Phi^*}(X'^2 + Y'^2) - \frac{1}{2}\eta\,\Omega(XY' - X'Y)\,.$$

[Φ^* denotes $\Phi(1+\varepsilon\Phi)$.] If a coordinate system, $x - y$, is introduced, inclined to $X - Y$ at a variable angle $\Theta(z)$, the term in $(XY' - X'Y)$ can be removed by selecting $\Theta' = \eta\,\Omega/2\sqrt{\Phi^*}$; $m^{(2)}$ becomes

$$m^{(2)} = -\left\{\frac{1+2\varepsilon\Phi}{8\sqrt{\Phi^*}}\Phi'' + \frac{\eta^2}{8}\frac{\Omega^2}{\sqrt{\Phi^*}}\right\}(x^2+y^2) + \\ + \left\{\left(\frac{1+2\varepsilon\Phi}{8\sqrt{\Phi^*}}D - \frac{1}{2}\eta Q\right)\cos 2\Theta + \right. \\ \left. + \left(\frac{1+2\varepsilon\Phi}{4\sqrt{\Phi^*}}P + \frac{1}{4}\eta\Delta\right)\sin 2\Theta\right\}(x^2 - y^2) + \\ + \left\{-\left(\frac{1+2\varepsilon\Phi}{4\sqrt{\Phi^*}}D - \eta Q\right)\sin 2\Theta + \right. \\ \left. + \left(\frac{1+2\varepsilon\Phi}{2\sqrt{\Phi^*}}P + \frac{1}{2}\eta\Delta\right)\cos 2\Theta\right\}xy + \\ + \frac{1}{2}\sqrt{\Phi^*}(x'^2 + y'^2)\,. \qquad (2.4)$$

The paraxial equations of motion are of the form

$$\frac{\mathrm{d}}{\mathrm{d}z}\left(\frac{\partial m^{(2)}}{\partial x'}\right)=\frac{\partial m^{(2)}}{\partial x}, \quad \frac{\mathrm{d}}{\mathrm{d}z}\left(\frac{\partial m^{(2)}}{\partial y'}\right)=\frac{\partial m^{(2)}}{\partial y} \tag{2.5}$$

and the system is orthogonal if these separate into an equation for x not containing y, and an equation for y not containing x. This will be the case if

$$\tan 2\Theta=\frac{\frac{1+2\varepsilon\Phi}{\sqrt{\Phi^*}}P+\eta\Delta}{\frac{1+2\varepsilon\Phi}{2\sqrt{\Phi^*}}D-2\eta Q}, \quad \text{with} \quad \Theta'=\frac{\eta}{2}\frac{\Omega}{\sqrt{\Phi^*}} \tag{2.6}$$

This is called the *orthogonality condition* [33, 69, 74, 216, 217]. Quadrupole systems for which this condition is not fulfilled are discussed by *Rose* [153a]; cf. *Carathéodory* [25].

This condition could in theory be satisfied in any of three ways (*Dušek* [45, 46]) but only one seems to be a practical possibility. Most generally, $\Theta(z)$ may indeed be a function of z, in which case electrodes and pole-pieces must be devised and constructed of such shapes that the condition is everywhere satisfied. A more reasonable solution is to set Θ equal to a constant, different from zero, giving $\Omega(z)=0$ and (non-relativistically), $P+\eta\Delta\sqrt{\Phi}\propto D-4\eta Q\sqrt{\Phi}$. An example of such a system, which we shall analyse in the section on field models, is discussed by *Dušek* [45]. Finally we may set Θ equal to 0 or $\pi/4$, so that either $P(z)$ and or $\Delta(z)$ do not vanish, while $D(z)$ and $Q(z)$ are zero everywhere or $D(z)$ and or $Q(z)$ do not vanish, while $P(z)=0$ and $\Delta(z)=0$ [again $\Omega(z)=0$]. With $\Theta=0$, therefore,

$$m^{(2)}=-\frac{1+2\varepsilon\Phi}{8\sqrt{\Phi^*}}\Phi''(x^2+y^2)+\left(\frac{1+2\varepsilon\Phi}{8\sqrt{\Phi^*}}D-\frac{1}{2}\eta Q\right)(x^2-y^2)+$$
$$+\frac{1}{2}\sqrt{\Phi^*}(x'^2+y'^2) \tag{2.7}$$

and

$$p=\frac{\partial m^{(2)}}{\partial x'}=\sqrt{\Phi^*}\,x'; \quad q=\frac{\partial m^{(2)}}{\partial y'}=\sqrt{\Phi^*}\,y'. \tag{2.8}$$

The physical appearance of a rectilinear system of which the paraxial properties are wholly described by $\Phi(z)$, $D(z)$ and $Q(z)$ can be deduced by writing the expressions for the electrostatic potential φ and the magnetic *scalar* potential φ^m ($\boldsymbol{B}=-\operatorname{grad}\varphi^m$) in terms of cylindrical polar co-ordinates, (r,θ,z):

$$\begin{aligned}
\varphi(r,\theta,z) &= \Phi-\frac{1}{4}\Phi''r^2+\frac{1}{64}\Phi^{(\mathrm{iv})}r^4-\cdots: && \varphi_0\text{-terms}\\
&+\cos 2\theta\left(\frac{1}{4}Dr^2-\frac{1}{48}D''r^4+\cdots\right): && \varphi_2\text{-terms}\\
&+\cos 4\theta\,(D_1r^4-\cdots)+\sin 4\theta\,(P_1r^4-\cdots): && \varphi_4\text{-terms} \quad (2.9)\\
\varphi^m(r,\theta,z) &= \sin 2\theta\left(\frac{1}{2}Qr^2-\frac{1}{24}Q''r^4+\cdots\right): && \varphi_2^m\text{-terms}\\
&+\cos 4\theta\,(\Delta_1r^4-\cdots)+\sin 4\theta\,(Q_1r^4-\cdots): && \varphi_4^m\text{-terms}.
\end{aligned}$$

The φ_0-terms correspond to a rotationally symmetric electric field, $E_z = -\Phi' + \cdots$, $E_r = \frac{1}{2}\Phi'' r - \cdots$; the φ_2-terms correspond to a system of electrodes possessing two planes of symmetry, the coordinate planes, since $\varphi_2(r, \theta, z) = \varphi_2(r, -\theta, z)$ and $\varphi_2(r, \pi/2 + \theta, z) = \varphi_2(r, \pi/2 - \theta, z)$; the φ_2^m-terms correspond to a system of pole-pieces possessing two planes of symmetry, mid-way between the coordinate planes, since

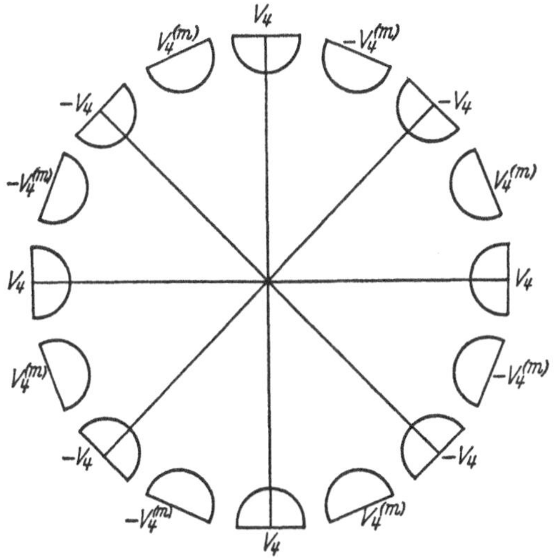

Fig. 2. The orientation of electrodes corresponding to φ_4 and φ_4^m when $P_1(z) = \Delta_1(z) \equiv 0$

$\varphi_2^m(r, \pi/4 + \theta^*, z) = \varphi_2^m(r, \pi/4 - \theta^*, z)$ and $\varphi_2^m(r, 3\pi/4 + \theta^*, z) = \varphi_2^m(r, 3\pi/4 - \theta^*, z)$ for any value of θ^*. The φ_4- and φ_4^m-terms in $\cos 4\theta$ correspond to systems of electrodes and pole-pieces possessing four planes of symmetry, the coordinate planes and the planes mid-way between them; the terms in $\sin 4\theta$ correspond to sets of electrodes or pole-pieces mid-way between the planes of symmetry of the $\cos 4\theta$-systems (see Fig. 2). Hence φ_0-terms describe round electrostatic lenses; φ_2- and φ_2^m-terms describe electrostatic and magnetic quadrupoles, so orientated that the coordinate planes intersect the electrodes and bisect the angle between the pole-pieces; φ_4- and φ_4^m-terms describe octopoles and if $P_1 = 0$ and $\Delta_1 = 0$, the coordinate planes again intersect the electrodes and pass mid-way between the pole-pieces.

To analyse the paraxial imagery of an orthogonal system for which $\Theta = \Delta = P = 0$, we employ the angle characteristic*. We suppose the

* The properties of orthogonal systems have been studied by *Larmor*, *Gullstrand* and *Herzberger*, and methodical accounts are to be found in *Carathéodory* [25], where these systems are treated as a special case of a more general class of systems, *Luneburg* [127] and *Herzberger* [98] where the most thorough discussion is to be found. (Cf. *Hawkes* [87, 89] for further references.) The present examination follows *Luneburg* closely; for additional details *Herzberger* should be consulted.

An account of quadrupoles as focusing devices is to be found in *Regenstreif* [151].

optic axis to be divided into three regions*, object space, lens space and image space, and we consider how the angle characteristic between a point in object space lying in the plane $z = z_O$ and a point in image space in $z = z_C$ is related to the angle characteristic between a second point in object space (x_o, y_o, z_o^*) and a second point in image space (x_c, y_c, z_c^*). We have

$$T_{oc} = T_{OC} + T_{oO} - T_{cC}$$

and in field-free space,

$$T_{oO} = V_{oO} + p_o x_o + q_o y_o - p_O x_O - q_O y_O$$
$$= (z_O - z_o^*) \{\sqrt{\varphi(1 + x'^2 + y'^2)} - p_o x' - q_o y'\}$$

in which we have used $p_o = p_O$, $q_o = q_O$; but $x' = p/\sqrt{\varphi - p_o^2 - q_o^2}$, $y' = q/\sqrt{\varphi - p_o^2 - q_o^2}$, so that $T_{oO} = (z_O - z_o^*)\sqrt{\varphi - p_o^2 - q_o^2}$. To the paraxial approximation, therefore,

$$T_{oc} = T_{OC} - z_o U_o \left(1 - \frac{p_o^2 + q_o^2}{2U_o^2}\right) + z_c U_c \left(1 - \frac{p_c^2 + q_c^2}{2U_c^2}\right)$$

in which $z_o = z_o^* - z_O$, $z_c = z_c^* - z_C$ and $U = \sqrt{\varphi}$.

Since the system has two planes of symmetry, p_o, p_c, q_o and q_c can only appear in the combinations p_o^2, q_o^2, p_c^2, q_c^2, $p_o p_c$ and $q_o q_c$. Writing

$$T_{OC} = \frac{1}{2}(A_x p_o^2 + A_y q_o^2 + C_x p_c^2 + C_y q_c^2) - (\bar{f}_x p_o p_c + \bar{f}_y q_o q_c)$$

we obtain from equations (2.2)

$$\begin{aligned} x_o &= (A_x + z_o/U_o)\, p_o - \bar{f}_x p_c; \quad x_c = -(C_x - z_c/U_c)\, p_c + \bar{f}_x p_o; \\ y_o &= (A_y + z_o/U_o)\, q_o - \bar{f}_y q_c; \quad y_c = -(C_y - z_c/U_c)\, q_c + \bar{f}_y q_o . \end{aligned} \tag{2.10}$$

If $p_c = 0$, x_o vanishes in the plane $z_o = -U_o A_x$ for all p_o, and if $p_o = 0$, x_c vanishes at $z_c = U_c C_x$ for all p_c; these two points are characteristic of the system, its x-foci in fact, and we write $z_{Fi}^{(x)} = U_c C_x$, $z_{Fo}^{(x)} = -U_o A_x$. Eliminating either p_c or p_o between the equations for x_o and x_c, we find that $\bar{f}_x x_c = (C_x - z_c/U_c)\, x_o$ or $\bar{f}_x x_o = (A_x + z_o/U_o)\, x_c$ if $U_o U_c \bar{f}_x^2 - (C_x U_c - z_c)(A_x U_o + z_o) = 0$, so that $x_c = x_o$ if $z_c = U_c(C_x - \bar{f}_x)$ and $z_o = U_o(\bar{f}_x - A_x)$; these are the x-principal points of the system, $z_{Hi}^{(x)} = U_c(C_x - \bar{f}_x) = z_{Fi}^{(x)} - U_c \bar{f}_x$ and $z_{Ho}^{(x)} = U_o(\bar{f}_x - A_x) = U_o \bar{f}_x + z_{Fo}^{(x)}$. Setting the origins of z_o and z_c at the object and image foci, we obtain $z_{Ho}^{(x)} = U_o \bar{f}_x$, $z_{Hi}^{(x)} = -U_c \bar{f}_x$ and the focal lengths are given by $f_{xo} = U_o \bar{f}_x$, $f_{xi} = -U_c \bar{f}_x$ and are connected by $f_{xo}/f_{xi} = -U_o/U_i$. With the same origins for z_o and z_c, we obtain Newton's lens formula, $z_o z_c = f_{xo} f_{xi}$ and the magnification, $M_x = x_c/x_o$, is given by $M_x = z_c/f_{xi} = f_{xo}/z_o$.

The pair of equations for y_o and y_c can be treated in exactly the same fashion, and y-foci, y-principal planes and a Newtonian lens formula, $z_o z_c = f_{yo} f_{yi}$, derived. The x- and y-cardinal points will not coincide, however, and if the hollow conical pencil defined by $x_o = y_o = 0$, p_o

* This apparent limitation can be removed by considering asymptotes.

$= r\cos\theta$, $q_o = r\sin\theta$ ($r =$ constant) is considered, we find

$$x_c = \bar{f}_x\left(1 - \frac{z_{xc}\,z_{xo}}{f_{xo}\,f_{xi}}\right) r\cos\theta$$

$$y_c = \bar{f}_y\left(1 - \frac{z_{yc}\,z_{yo}}{f_{yo}\,f_{yi}}\right) r\sin\theta$$

($z_{xc} \ldots z_{yo}$ denote the abscissae of the object and current planes with respect to the relevant foci). The beam therefore becomes elliptical in cross-section and collapses to a line image at the two planes for which z_o and z_c satisfy Newton's lens formulae.

For a particular value of z_c, ζ_c say, the beam is circular, and from the relations

$$\bar{f}_x x_c = \{\bar{f}_x^2 - (C_x - z_c/U_c)\,(A_x + z_o/U_o)\}\, p_o$$

$$\bar{f}_y y_c = \{\bar{f}_y^2 - (C_y - z_c/U_c)\,(A_y + z_o/U_o)\}\, q_o$$

we deduce that

$$\bar{f}_x - \frac{\bar{A}_x}{\bar{f}_x}\left(C_x - \frac{\zeta_c}{U_c}\right) = -\bar{f}_y + \frac{\bar{A}_y}{\bar{f}_y}\left(C_y - \frac{\zeta_c}{U_c}\right)$$

(where $\bar{A} = A + z_o/U_o$) or, since $M_x = \bar{f}_x/\bar{A}_x$, $M_y = \bar{f}_y/\bar{A}_y$,

$$\frac{\zeta_c}{U_c}\left(\frac{1}{M_x} + \frac{1}{M_y}\right) = \left(\frac{C_x}{M_x} - \bar{f}_x\right) + \left(\frac{C_y}{M_y} - \bar{f}_y\right).$$

With ζ_c as origin of coordinates, we write

$$C_x/M_x - \bar{f}_x = -(C_y/M_y - \bar{f}_y) = \lambda$$

and so equations (2.10) become

$$x_c = \bar{f}_x p_o - (M_x \bar{f}_x - z_c/U_c + M_x\lambda)\, p_c$$

$$y_c = \bar{f}_y q_o - (M_y \bar{f}_y - z_c/U_c - M_y\lambda)\, q_c$$

$$x_o = (\bar{f}_x/M_x)\, p_o - \bar{f}_x p_c$$

$$y_o = (\bar{f}_y/M_y)\, q_o - \bar{f}_y q_c$$

or

$$x_c = M_x x_o + (z_c/U_c - M_x\lambda)\, p_c$$

$$y_c = M_y y_o + (z_c/U_c + M_y\lambda)\, q_c\,.$$

The astigmatic difference is thus $U_c(M_x + M_y)\,\lambda$. If the magnifications corresponding to $z_o = 0$ are $M_x^{(o)}$, $M_y^{(o)}$, those corresponding to arbitrary z_o will be

$$M_x = \frac{\bar{f}_x}{A + z_o/U_o} = \frac{M_x^{(o)}}{1 + M_x^{(o)} z_o/\bar{f}_x U_c}, \quad M_y = \frac{M_y^{(o)}}{1 + M_y^{(o)} z_o/\bar{f}_y U_c}.$$

Similarly, if the astigmatic difference corresponding to $z_o = 0$ is Λ_0, the general expression will be

$$\Lambda = \Lambda_0 + \frac{z_o}{2U_o}\,(M_y M_y^{(o)} - M_x M_x^{(o)})\,.$$

The paraxial properties of a system of quadrupole lenses and electrostatic round lenses are thus wholly defined by the positions of the foci and principal points for the x z-plane and the y z-plane. Generally,

$$\begin{pmatrix} x_c \\ p_c \end{pmatrix} = \begin{pmatrix} z_o^* & \bar{f}_x(1 - z_o^* z_c^*) \\ -1/\bar{f}_x & z_o^* \end{pmatrix} \begin{pmatrix} x_o \\ p_o \end{pmatrix}$$

$$\begin{pmatrix} y_c \\ q_c \end{pmatrix} = \begin{pmatrix} z_o^* & \bar{f}_y(1 - z_o^* z_c^*) \\ -1/\bar{f}_y & z_o^* \end{pmatrix} \begin{pmatrix} y_o \\ q_o \end{pmatrix}$$

in which $z_c^* = (C_x - z_c/U_c)/\bar{f}_x$ and $z_o^* = (A_x + z_o/U_o)/\bar{f}_x$. The determinants of these *transfer matrices* are both unity. A more convenient form of these matrices is the following:

$$\begin{pmatrix} x_c \\ p_c \end{pmatrix} = \begin{pmatrix} \dfrac{z_c - z_{Fi}^{(x)}}{f_{xi}} & \dfrac{f_{xi}}{U_c}\left\{\dfrac{(z_c - z_{Fi}^{(x)})}{f_{xi}} \dfrac{(z_o - z_{Fo}^{(x)})}{f_{xo}} - 1\right\} \\ \dfrac{U_c}{f_{xi}} & \dfrac{z_o - z_{Fo}^{(x)}}{f_{xo}} \end{pmatrix} \begin{pmatrix} x_o \\ p_o \end{pmatrix} \tag{2.11}$$

with a similar expression for the y z-plane.

Practical quadrupoles display features that have led to the use of other transfer matrices than these, however. The potential distributions $D(z)$, $Q(z)$ approach zero asymptotically, so that asymptotic cardinal elements may have to be defined; or the distributions may be assumed to vanish outside certain limits, $z = \pm\zeta$. In the first case, it is natural to relate the slope and height of the incident asymptote in some plane to the slope and height of the emergent asymptote in the same plane (the symmetry plane of D, Q is the normal choice, when one exists). In the second case, $x(\zeta)$, $y(\zeta)$, $p(\zeta)$ and $q(\zeta)$ are related to $x(-\zeta)$, $y(-\zeta)$, $p(-\zeta)$ and $q(-\zeta)$.

The coordination between asymptotes is employed by *Dušek* [45, 46], whose analysis we shall examine in detail. First, however, we set out the coordination between the slope of the straight line defined by x_o, y_o, p_o and q_o and its point of intersection with the symmetry plane of a field distribution symmetrical about $z = 0$, and the slope and point of intersection of the line defined by x_c, y_c, p_c and q_c, when $U_o = U_c = U$; thus if these are the asymptotic values of the coordinates and momenta, the connection between Dušek's asymptotic coordination and the present transfer matrices will be obvious.

The incident asymptote is written $\mathfrak{x} = \mathfrak{M}_- z + \mathfrak{N}_-$ and the emergent asymptote, $\mathfrak{x} = \mathfrak{M}_+ z + \mathfrak{N}_+$; $\mathfrak{x}$ denotes either x or y and $\mathfrak{M}$, $\mathfrak{N}$ take different values in each case. These asymptotes may be written $\mathfrak{x} = \mathfrak{p}_o z/U + \mathfrak{N}_-$, $\mathfrak{x} = \mathfrak{p}_c z/U + \mathfrak{N}_+$ (in which $\mathfrak{p}$ denotes p or q) or, considering the foci, $\mathfrak{x} = \mathfrak{p}_o z/U + (\mathfrak{x}_{Fo} - \mathfrak{p}_o z_{Fo}/U)$, $\mathfrak{x} = \mathfrak{p}_c z/U + (\mathfrak{x}_{Fi} - \mathfrak{p}_c z_{Fi}/U)$; using $\mathfrak{p}_c = -\mathfrak{x}_{Fo}/\bar{f}$ and $\mathfrak{x}_{Fi} = \bar{f}\mathfrak{p}_o$, we obtain

$$\begin{pmatrix} \mathfrak{M}_+ \\ \mathfrak{N}_+ \end{pmatrix} = \begin{pmatrix} -\dfrac{z_{Fo}}{U\bar{f}} & -\dfrac{1}{U\bar{f}} \\ \dfrac{z_{Fi} z_{Fo}}{U\bar{f}} + \bar{f}U & \dfrac{z_{Fi}}{U\bar{f}} \end{pmatrix} \begin{pmatrix} \mathfrak{M}_- \\ \mathfrak{N}_- \end{pmatrix}.$$

Substituting $U\bar{f} = f_o = -f_i$, $z_{Fo} = -z_{Fi} = z_F$, the transfer matrix becomes

$$\begin{pmatrix} -\dfrac{z_F}{f_o} & -\dfrac{1}{f_o} \\ -\dfrac{z_F^2}{f_o} + f_o & -\dfrac{z_F}{f_o} \end{pmatrix} \tag{2.12}$$

the determinant of which is again equal to unity. [This is the matrix which *Dušek* denotes $\begin{pmatrix} \overset{i}{\nu} & \overset{i}{\varrho} \\ \overset{i}{\tau} & \overset{i}{\nu} \end{pmatrix}$ for the i-th element of a system.]

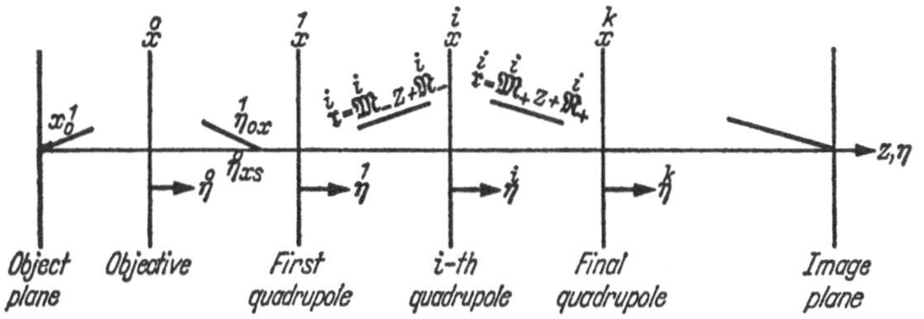

Fig. 3. An orthogonal system consisting of objective lens and quadrupoles

We now consider a potential model which is symmetrical about the plane $z = 0$ and falls to zero at $z = \pm\zeta$. Using the same techniques, we find

$$\begin{pmatrix} \mathfrak{x}_\zeta \\ \mathfrak{p}_\zeta \end{pmatrix} = \begin{pmatrix} -\dfrac{\zeta + z_F}{f_o} & \dfrac{f_o}{U} - \dfrac{(\zeta + z_F)^2}{U f_o} \\ -\dfrac{U}{f_o} & -\dfrac{\zeta + z_F}{f_o} \end{pmatrix} \begin{pmatrix} \mathfrak{x}_{-\zeta} \\ \mathfrak{p}_{-\zeta} \end{pmatrix} \tag{2.13}$$

and again, the determinant is unity.

The transfer matrix relating the asymptote emerging from one element, $(\overset{i}{\mathfrak{M}}_+, \overset{i}{\mathfrak{N}}_+)$, to the incident asymptote at the next, $(\overset{i+1}{\mathfrak{M}}_-, \overset{i+1}{\mathfrak{N}}_-)$, distant D_i, is clearly

$$\begin{pmatrix} \overset{i+1}{\mathfrak{M}}_- \\ \overset{i+1}{\mathfrak{N}}_- \end{pmatrix} = \begin{pmatrix} 1 & 0 \\ D_i & 1 \end{pmatrix} \begin{pmatrix} \overset{i}{\mathfrak{M}}_+ \\ \overset{i}{\mathfrak{N}}_+ \end{pmatrix}; \tag{2.14a}$$

if the drift space between two finite models is of length D_i, then

$$\begin{pmatrix} \overset{i+1}{\mathfrak{x}}_{-\zeta} \\ \overset{i+1}{\mathfrak{p}}_o \end{pmatrix} = \begin{pmatrix} 1 & D_i/U \\ 0 & 1 \end{pmatrix} \begin{pmatrix} \overset{i}{\mathfrak{x}}_\zeta \\ \overset{i}{\mathfrak{p}}_c \end{pmatrix}. \tag{2.14b}$$

The imagery of a succession of elements is conveniently analysed in Dušek's notation. Suppose that the system consists of an objective lens (not necessarily round) and k elements (see Fig. 3). The final asymptote is related to the asymptote entering the first of the k elements as follows:

$$\begin{pmatrix} \overset{k}{\mathfrak{M}}_+ \\ \overset{k}{\mathfrak{N}}_+ \end{pmatrix} = \begin{pmatrix} \overset{k}{\nu} & \overset{k}{\varrho} \\ \overset{k}{\tau} & \overset{k}{\nu} \end{pmatrix} \begin{pmatrix} 1 & D_{k-1} \\ 0 & 1 \end{pmatrix} \begin{pmatrix} \overset{k-1}{\nu} & \overset{k-1}{\varrho} \\ \overset{k-1}{\tau} & \overset{k-1}{\nu} \end{pmatrix} \begin{pmatrix} 1 & D_{k-2} \\ 0 & 1 \end{pmatrix} \cdots \begin{pmatrix} \overset{1}{\mathfrak{M}}_- \\ \overset{1}{\mathfrak{N}}_- \end{pmatrix}$$

$$= \begin{pmatrix} \mathrm{I} & \mathrm{II} \\ \mathrm{III} & \mathrm{IV} \end{pmatrix} \begin{pmatrix} \overset{1}{\mathfrak{M}}_- \\ \overset{1}{\mathfrak{N}}_- \end{pmatrix}$$

If the object lies on the axis, the asymptote emerging from the objective, which is symmetrical about $\overset{0}{z}$, is of the form

$$x = \frac{x_o'}{m_x}\left(\overset{0}{z} - \overset{0}{z}_{xs}\right) \qquad y = \frac{y_o'}{m_y}\left\{\overset{0}{z} - \left(\overset{0}{z}_{ys} - \varLambda\right)\right\}$$

where m_x, m_y are the objective magnifications, and $\varLambda$ is the astigmatic difference. If the first orthogonal system is at distance D_o, the x and y expressions for $\overset{1}{\mathfrak{M}}_-$ and $\overset{1}{\mathfrak{N}}_-$ are thus

$$x\colon\ \overset{1}{\mathfrak{M}}_- = \frac{x_o'}{m_x},\ \overset{1}{\mathfrak{N}}_- = -\frac{x_o'}{m_x}(z_{xs} - D_o)\,,$$

$$y\colon\ \overset{1}{\mathfrak{M}}_- = \frac{y_o'}{m_y},\ \overset{1}{\mathfrak{N}}_- = -\frac{y_o'}{m_y}(z_{xs} - D_o - \varLambda)\,.$$

The terminal asymptotes become

$$x = \frac{x_o'}{m_x}\left[\{\mathrm{I}_x - \mathrm{II}_x(z_{xs} - D_o)\}\, z + \{\mathrm{III}_x - \mathrm{IV}_x(z_{xs} - D_o)\}\right],$$

$$y = \frac{y_o'}{m_y}\left[\{\mathrm{I}_y - \mathrm{II}_y(z_{xs} - D_o - \varLambda)\}\, z + \{\mathrm{III}_y - \mathrm{IV}_y(z_{xs} - D_0 - \varLambda)\}\right].$$

If the complete system is required to be stigmatic, the points at which these rays cross the axis must coincide, at $z = \zeta$ say:

$$\begin{aligned} &\bar{z}\zeta - \frac{\mathrm{I}_x}{\mathrm{II}_x}\zeta + \frac{\mathrm{IV}_x}{\mathrm{II}_x}\bar{z} - \frac{\mathrm{III}_x}{\mathrm{II}_x} = 0 \\ &\bar{z}\zeta - \left(\frac{\mathrm{I}_y}{\mathrm{II}_y} + \varLambda\right)\zeta + \frac{\mathrm{IV}_y}{\mathrm{II}_y}\bar{z} - \left(\frac{\mathrm{III}_y}{\mathrm{II}_y} + \frac{\mathrm{IV}_y}{\mathrm{II}_y}\varLambda\right) = 0 \end{aligned} \tag{2.15}$$

($\bar{z}$ denotes $z_{xs} - D_o$). If, in addition, it is to be orthomorphic, the final magnifications, M_x and M_y, must be equal in magnitude (though not necessarily in sign). The slopes of the terminal asymptotes are x_o'/M_x and y'/M_y, so that

$$\begin{aligned} \frac{1}{M_x} &= \frac{1}{m_x}(\mathrm{I}_x - \mathrm{II}_x\,\bar{z}) \\ \frac{1}{M_y} &= \frac{1}{m_y}\{\mathrm{I}_y - \mathrm{II}_y(\bar{z} - \varLambda)\}\,. \end{aligned} \tag{2.16}$$

The system will be stigmatic for all positions of the object plane only if there exists a solution for ζ of equations (2.15) for all values of $\bar{z}$ ($\varLambda$ varies with $\bar{z}$). *Dušek* considers a slightly different problem: under what conditions will the elements 1 to k produce a stigmatic and orthomorphic image if Λ, m_x and m_y are fixed but $\bar{z}$ is variable? The conditions for stigmatism and equality of final magnifications imply

$$\frac{\mathrm{I}_x}{\mathrm{II}_x} = \frac{\mathrm{I}_y}{\mathrm{II}_y} + \Lambda \qquad \frac{\mathrm{III}_x}{\mathrm{II}_x} = \frac{\mathrm{III}_y}{\mathrm{II}_y} + \Lambda \frac{\mathrm{IV}_y}{\mathrm{II}_y}$$

$$\frac{\mathrm{IV}_x}{\mathrm{II}_x} = \frac{\mathrm{IV}_y}{\mathrm{II}_y} \qquad \frac{\mathrm{II}_x}{m_x} = \pm \frac{\mathrm{II}_y}{m_y}\,.$$

The three equations not containing m_x, m_y can be shown to lead to

$$\mathrm{II}_x = \pm \mathrm{II}_y$$

so that only if $m_x = m_y$ are the conditions not self-contradictory. If however the objective is rotationally symmetric ($\Lambda = 0$ and $m_x = m_y$), the conditions become $\mathrm{I}_x = \mathrm{I}_y$, $\mathrm{II}_x = \mathrm{II}_y$, $\mathrm{III}_x = \mathrm{III}_y$ and $\mathrm{IV}_x = \mathrm{IV}_y$: for a set of orthogonal systems to produce a stigmatic orthomorphic image of any object plane, the x- and y-cardinal elements of the overall system must coincide.

The equations of motion and their asymptotic solutions

The paraxial equations of motion in rectilinear orthogonal systems for which $\Theta = 0$ and $\Delta(z)$ and $P(z)$ vanish (twist-free systems) are of the form

$$p' = \partial m^{(2)}/\partial x \quad q' = \partial m^{(2)}/\partial y$$

with $p = \sqrt{\Phi^*}\, x'$, $q = \sqrt{\Phi^*}\, y'$, so that

$$\begin{aligned} &\frac{\mathrm{d}}{\mathrm{d}z}(\sqrt{\Phi^*}\, x') + \frac{(1 + 2\,\varepsilon\,\Phi)\,\Phi'' - (1 + 2\,\varepsilon\,\Phi)\,D + 4\,\eta\,Q\sqrt{\Phi^*}}{4\sqrt{\Phi^*}}\,x = 0 \\ &\frac{\mathrm{d}}{\mathrm{d}z}(\sqrt{\Phi^*}\, y') + \frac{(1 + 2\,\varepsilon\,\Phi)\,\Phi'' + (1 + 2\,\varepsilon\,\Phi)\,D - 4\,\eta\,Q\sqrt{\Phi^*}}{4\sqrt{\Phi^*}}\,y = 0 \end{aligned} \tag{2.17}$$

or, non-relativistically,

$$\begin{aligned} &x'' + \frac{\Phi'}{2\Phi}x + \frac{\Phi'' - D + 4\,\eta\,Q\sqrt{\Phi}}{4\Phi}\,x = 0 \\ &y'' + \frac{\Phi'}{2\Phi}y + \frac{\Phi'' + D - 4\,\eta\,Q\sqrt{\Phi}}{4\Phi}\,y = 0\,. \end{aligned} \tag{2.18}$$

As with round electrostatic lenses, we may eliminate the second derivative of $\Phi(z)$ by writing $x = X\,\Phi^{*\,-1/4}$, $y = Y\Phi^{*\,-1/4}$ (or $x = X\,\Phi^{-1/4}$ $y = Y\,\Phi^{-1/4}$ when terms in ε are neglected). Non-relativistically, we obtain

$$\begin{aligned} &X'' + \left(\frac{3}{16}\frac{\Phi'^2}{\Phi^2} - \frac{D - 4\,\eta\,Q\sqrt{\Phi}}{4\Phi}\right)X = 0 \\ &Y'' + \left(\frac{3}{16}\frac{\Phi'^2}{\Phi^2} + \frac{D - 4\,\eta\,Q\sqrt{\Phi}}{4\Phi}\right)Y = 0\,. \end{aligned} \tag{2.19}$$

That asymptotes always exist for twist-free systems is shown by *Dušek* [45], following the reasoning of *Glaser* and *Bergmann* [70]. We assume that this has been established, and deduce the expressions for the asymptotes from the equations of motion. We write

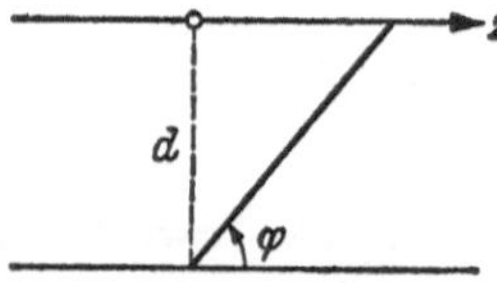

Fig. 4. The auxiliary variable $\varphi (z = d \cot \varphi)$

$$\mathfrak{x} = \mathfrak{X}(\varphi) \operatorname{cosec} \varphi$$
$$= \{C_1 \mathfrak{X}_1(\varphi) + C_2 \mathfrak{X}_2(\varphi)\} \operatorname{cosec} \varphi$$

in which $\mathfrak{x}$, $\mathfrak{X}$ denote either x or y. As usual, $z = 0$ is the symmetry plane of an element, and $z = d \cot \varphi$ (see Fig. 4); the object asymptote corresponds to $\varphi = \pi$, and the image asymptote to $\varphi = 0$. The scale factor d can always be eliminated by writing $\eta = z/d$.

Then

$$\frac{\mathrm{d}\mathfrak{x}}{\mathrm{d}\eta} = \mathfrak{X} \cos \varphi - \frac{\mathrm{d}\mathfrak{X}}{\mathrm{d}\varphi} \sin \varphi$$

$$\mathfrak{x} - \eta \frac{\mathrm{d}\mathfrak{x}}{\mathrm{d}\eta} = \mathfrak{X} \sin \varphi + \frac{\mathrm{d}\mathfrak{X}}{\mathrm{d}\varphi} \cos \varphi .$$

The object asymptote is $\mathfrak{x} = \mathfrak{M}_- z + \mathfrak{N}_-$, where

$$\mathfrak{M}_- = \frac{1}{d} \operatorname*{Lt}_{\varphi \to \pi} \left(\frac{\mathrm{d}\mathfrak{x}}{\mathrm{d}\eta}\right) \quad \text{and} \quad \mathfrak{N}_- = \operatorname*{Lt}_{\varphi \to \pi} \left(\mathfrak{x} - \cot \varphi \frac{\mathrm{d}\mathfrak{x}}{\mathrm{d}\eta}\right)$$

or

$$\begin{aligned} \mathfrak{M}_- &= -\mathfrak{X}_\pi/d = -(C_1 \mathfrak{X}_{1\pi} + C_2 \mathfrak{X}_{2\pi})/d \\ \mathfrak{N}_- &= -\left(\frac{\mathrm{d}\mathfrak{X}}{\mathrm{d}\varphi}\right)_\pi = -\left\{C_1 \left(\frac{\mathrm{d}\mathfrak{X}_1}{\mathrm{d}\varphi}\right)_\pi + C_2 \left(\frac{\mathrm{d}\mathfrak{X}_2}{\mathrm{d}\varphi}\right)_\pi\right\}. \end{aligned} \tag{2.20a}$$

Likewise, the image asymptote, $\mathfrak{x} = \mathfrak{M}_+ z + \mathfrak{N}_+$, is defined by

$$\begin{aligned} \mathfrak{M}_+ &= \mathfrak{X}_0/d = (C_1 \mathfrak{X}_{10} + C_2 \mathfrak{X}_{20})/d \\ \mathfrak{N}_+ &= \left(\frac{\mathrm{d}\mathfrak{X}}{\mathrm{d}\varphi}\right)_0 = C_1 \left(\frac{\mathrm{d}\mathfrak{X}_1}{\mathrm{d}\varphi}\right)_0 + C_2 \left(\frac{\mathrm{d}\mathfrak{X}_2}{\mathrm{d}\varphi}\right)_0. \end{aligned} \tag{2.20b}$$

Denoting $\mathrm{d}\,\mathfrak{X}/\mathrm{d}\varphi$ by $\dot{\mathfrak{X}}$, we have

$$d\mathfrak{M}_- = -C_1 \mathfrak{X}_{1\pi} - C_2 \mathfrak{X}_{2\pi} \quad d\mathfrak{M}_+ = C_1 \mathfrak{X}_{10} + C_2 \mathfrak{X}_{20}$$
$$\mathfrak{N}_- = -C_1 \dot{\mathfrak{X}}_{1\pi} - C_2 \dot{\mathfrak{X}}_{2\pi} \quad \mathfrak{N}_+ = C_1 \dot{\mathfrak{X}}_{10} + C_2 \dot{\mathfrak{X}}_{20}$$

and eliminating C_1 and C_2,

$$\begin{aligned} d\mathfrak{M}_+ &= \varkappa (d\mathfrak{M}_- \Omega^{1\,0}_{\pi\,0} + \mathfrak{N}_- \Omega_{0\pi}) \\ \mathfrak{N}_+ &= \varkappa (d\mathfrak{M}_- \Omega^{1\,1}_{\pi\,0} + \mathfrak{N}_- \Omega^{1\,0}_{0\,\pi}) \end{aligned} \tag{2.21}$$

in which

$$\frac{1}{\varkappa} = \begin{vmatrix} \mathfrak{X}_{1\pi} & \mathfrak{X}_{2\pi} \\ \dot{\mathfrak{X}}_{1\pi} & \dot{\mathfrak{X}}_{2\pi} \end{vmatrix}$$

$$\Omega_{mn} = -\Omega_{nm} = \begin{vmatrix} \mathfrak{X}_{1m} & \mathfrak{X}_{2m} \\ \mathfrak{X}_{1n} & \mathfrak{X}_{2n} \end{vmatrix} \quad (\Omega_{mm} = \Omega_{nn} = 0)$$

$$\Omega^{10}_{mn} = -\Omega^{01}_{nm} = \begin{vmatrix} \dot{\mathfrak{X}}_{1m} & \dot{\mathfrak{X}}_{2m} \\ \mathfrak{X}_{1n} & \mathfrak{X}_{2n} \end{vmatrix}$$

$$\Omega^{01}_{mn} = -\Omega^{10}_{nm} = \begin{vmatrix} \mathfrak{X}_{1m} & \mathfrak{X}_{2m} \\ \dot{\mathfrak{X}}_{1n} & \dot{\mathfrak{X}}_{2n} \end{vmatrix}$$

$$\Omega^{11}_{mn} = -\Omega^{11}_{nm} = \begin{vmatrix} \dot{\mathfrak{X}}_{1m} & \dot{\mathfrak{X}}_{2m} \\ \dot{\mathfrak{X}}_{1n} & \dot{\mathfrak{X}}_{2n} \end{vmatrix} \quad (\Omega^{11}_{mm} = \Omega^{11}_{nn} = 0)\,.$$

The inverse of equations (2.21) is

$$\begin{aligned} d\mathfrak{M}_- &= \varkappa\,(d\mathfrak{M}_+\,\Omega^{10}_{0\pi} - \mathfrak{N}_+\,\Omega_{0\pi}) \\ \mathfrak{N}_- &= \varkappa\,(-d\mathfrak{M}_+\,\Omega^{11}_{0\pi} + \mathfrak{N}_+\,\Omega^{10}_{\pi 0}) \end{aligned} \qquad (2.22)$$

and it can be shown that $\Omega^{10}_{\pi 0} = \Omega^{10}_{0\pi}$; furthermore, the three constants $\varkappa\,\Omega_{0\pi}$, $\varkappa\,\Omega^{11}_{\pi 0}$ and $\varkappa\,\Omega^{10}_{\pi 0}$ are interrelated:

$$(\Omega^{10}_{\pi 0})^2 - \Omega_{0\pi}\,\Omega^{11}_{\pi 0} = 1/\varkappa^2\,.$$

From equation (2.12), it is clear that the asymptotic cardinal elements are given by

$$-f_o = f_i = d/\varkappa\,\Omega_{0\pi}$$

$$z_{Fo} = -z_{Fi} = d\,\frac{\Omega^{10}_{\pi 0}}{\Omega_{0\pi}}\,.$$

3. Primary aberrations

By "primary" aberrations are meant all the types of defect that can be investigated by means of first-order perturbation theory: third-order geometrical aberrations, paraxial chromatic aberrations, certain mechanical aberrations, space charge aberrations and, if the accelerating voltage is not too high and the non-relativistic theory is employed, relativistic "aberrations". These several defects have been investigated with varying degrees of thoroughness: the geometrical and chromatic aberrations have been very elaborately investigated; some work has been done on the mechanical aberrations; the space charge aberrations can usually be disregarded except in regions where geometrical optics is itself unsatisfactory; the effect of relativity is normally either dismissed as insignificant ($\varepsilon = 0$) or included in the formulae from the outset, and not calculated by means of perturbation theory.

Any first-order perturbation alters the refractive index function m from $m = m^{(2)}$, the group of terms that yields paraxial imagery, to $m = m^{(2)} + m^{(P)}$. If, for example, the geometrical aberrations are to be calculated, $m^P = m^{(4)}$, while for the chromatic aberrations, $m^{(P)} = \partial m^{(2)}/\partial \Phi_0$. Aberration coefficients for each type of defect can be established in a number of ways. In the "trajectory method", the *Euler* equations of $\delta \int (m^{(2)} + m^{(P)})\,dz$ are obtained, and solutions sought of the form $x(z) = x^{(1)}(z) + x^P(z)$ in which $x^{(1)}(z)$ is the paraxial solution, satisfying the *Euler* equations of $\delta \int m^{(2)}\,dz = 0$. This leads to inhomogeneous differential equations, but since the homogeneous parts of these equations are identical with the paraxial equations of motion, the inhomogeneous equations can be readily solved by variation of parameters. In the method of perturbation characteristic functions, due to *Glaser* [71] and perfected by *Sturrock* [214], perturbed forms of one of Hamilton's characteristic functions are obtained, and partial differentiation leads directly to the aberration coefficients. A third method, exploited by *Meads* [130], yields solutions of the perturbed equations of motion in the form of integral equations; forty third-order geometrical aberration coefficients are obtained, connected by twenty-four relationships, and since the object of *Meads'* work was to develop a computer programme to calculate the coefficients, this redundancy provided a convenient check of the correctness of the computations.

Common to both the trajectory method and *Meads'* technique using integral equations is the need to derive explicitly the third-order equations of motion. The general form of these is to be found only in *Melkich* [133], and special cases of his equations are derived (from first principles) by *Scherzer* [156], *Bernard* and *Hue* [15, 16], *Grivet* and *Septier* [81, 82] (or see *Septier* [182]), *Markovich* and *Tsukkerman* [129], *Strashkevich* [210] and *Meads* [130] *i. a.*; expressions for the equations of motion in the completely general situation of a system with arbitrarily curved optic axis have also been derived, and the equations for rectilinear systems can be deduced as a special case of these. The best texts are probably *Sturrock* [216] and *Vandakurov* [288], for although the latter is not the earliest attempt to derive these general equations, *Vandakurov* noticed that some of his predecessors' work contained errors, arising from a faulty calculation of second-order effects. The special case of electrostatic systems is mentioned by *Strashkevich* [206]. The relevant earlier works are [76, 77, 78, 104–107, 146, 199, 200, 204, 223].

The equations of motion have the following general form:

$$x'' + \frac{1}{2}\frac{\Phi'}{\Phi}x' + \frac{\Phi'' - D + 4\eta Q\sqrt{\Phi}}{4\Phi}x$$

$$= \left\{-\frac{(\Phi''-D)^2}{16\Phi^2} + \frac{1}{32}\frac{\Phi^{(\mathrm{iv})}}{\Phi} - \frac{1}{24}\frac{D''}{\Phi} - \frac{1}{8}\eta Q\frac{\Phi''-D}{\Phi^{3/2}} + \frac{1}{12}\eta\frac{Q''}{\sqrt{\Phi}} + \right.$$

$$\left. + 2\frac{D_1}{\Phi} - 4\eta\frac{Q_1}{\sqrt{\Phi}}\right\}x^3 +$$

$$+ \left\{-\frac{(\Phi''-D)(\Phi''+D)}{16\Phi^2} + \frac{1}{32}\frac{\Phi^{(\mathrm{iv})}}{\Phi} - \frac{1}{8}\eta Q\frac{\Phi''+D}{\Phi^{3/2}} + \frac{1}{4}\eta\frac{Q''}{\sqrt{\Phi}} - \right.$$

$$\left. - 6\frac{D_1}{\Phi} + 12\eta\frac{Q_1}{\sqrt{\Phi}}\right\}x y^2 +$$

$$+\left\{\frac{\Phi'''-D'}{8\Phi}-\frac{\Phi'(\Phi''-D)}{8\Phi^2}\right\}x^2 x'+\left\{\frac{\Phi'''+D'}{8\Phi}-\frac{\Phi'(\Phi''+D)}{8\Phi^2}\right\}y^2 x'+$$

$$+\eta\frac{Q'}{\sqrt{\Phi}}x y y'-\frac{\Phi''-D+6\eta Q\sqrt{\Phi}}{4\Phi}x x'^2-$$

$$-\frac{\Phi''-D+2\eta Q\sqrt{\Phi}}{4\Phi}x y'^2+\eta\frac{Q}{\sqrt{\Phi}}y x' y'-$$

$$-\frac{1}{2}\frac{\Phi'}{\Phi}x'(x'^2+y'^2) \qquad (3.1\,\mathrm{a})$$

and

$$y''+\frac{\Phi'}{2\Phi}y'+\frac{\Phi''+D-4\eta Q\sqrt{\Phi}}{4\Phi}y$$

$$=\left\{-\frac{(\Phi''+D)^2}{16\Phi^2}+\frac{1}{32}\frac{\Phi^{(iv)}}{\Phi}+\frac{1}{24}\frac{D''}{\Phi}+\frac{1}{8}\eta Q\frac{\Phi''+D}{\Phi^{3/2}}-\frac{1}{12}\eta\frac{Q''}{\sqrt{\Phi}}+\right.$$

$$\left.+2\frac{D_1}{\Phi}-4\eta\frac{Q_1}{\sqrt{\Phi}}\right\}y^3+$$

$$+\left\{-\frac{(\Phi''-D)(\Phi''+D)}{16\Phi^2}+\frac{1}{32}\frac{\Phi^{(iv)}}{\Phi}+\frac{1}{8}\eta Q\frac{\Phi''-D}{\Phi^{3/2}}-\frac{1}{4}\eta\frac{Q''}{\sqrt{\Phi}}-\right.$$

$$\left.-6\frac{D_1}{\Phi}+12\eta\frac{Q_1}{\sqrt{\Phi}}\right\}x^2 y$$

$$+\left\{\frac{\Phi'''-D'}{8\Phi}-\frac{\Phi'(\Phi''-D)}{8\Phi^2}\right\}x^2 y'+\left\{\frac{\Phi'''+D'}{8\Phi}-\frac{\Phi'(\Phi''+D)}{8\Phi^2}\right\}y^2 y'-$$

$$-\eta\frac{Q'}{\sqrt{\Phi}}x y x'-$$

$$-\frac{\Phi''+D-2\eta Q\sqrt{\Phi}}{4\Phi}y x'^2-\frac{\Phi''+D-6\eta Q\sqrt{\Phi}}{4\Phi}y y'^2-$$

$$-\eta\frac{Q}{\sqrt{\Phi}}x x' y'-\frac{\Phi'}{2\Phi}(x'^2+y'^2)y'. \qquad (3.1\,\mathrm{b})$$

3.1 Variation of parameters

Equations (3.1) are solved by writing

$$x=x^{(1)}+x^{(3)}, \quad y=y^{(1)}+y^{(3)}$$

in which $x^{(1)}$ and $y^{(1)}$ are the solutions of the paraxial equations of motion, obtained by neglecting the right-hand sides of the equations. This yields differential equations for $x^{(3)}$ and $y^{(3)}$, the left-hand sides of which are the same as those of the paraxial equations; the inhomogeneous differential equations can thus be solved straightforwardly by the method of variation of parameters. The detailed solution depends upon the boundary conditions (cf. *Hawkes* [93]), which in turn are related to the type of paraxial solutions that are appropriate. If the latter are defined in terms of the position and slope of a ray in the object plane, we write

$$\begin{aligned} x^{(1)}(z) &= x_o\, s_x(z) + p_o\, t_x(z) \\ y^{(1)}(z) &= y_o\, s_y(z) + q_o\, t_y(z)\,. \end{aligned} \tag{3.2}$$

If the positions of the ray in two planes, the object and aperture planes, are more suitable, we write

$$\begin{aligned} x^{(1)}(z) &= x_o\, g_x(z) + x_a\, h_x(z) \\ y^{(1)}(z) &= y_o\, g_y(z) + y_a\, h_y(z)\,. \end{aligned} \tag{3.3}$$

If the slopes of the ray in two planes are chosen, we have

$$\begin{aligned} x^{(1)}(z) &= p_o\, \sigma_x(z) + p_a\, \tau_x(z) \\ y^{(1)}(z) &= q_o\, \sigma_y(z) + q_a\, \tau_y(z)\,. \end{aligned} \tag{3.4}$$

The corresponding boundary conditions are

$$\begin{aligned} s_{xo} = s_{yo} = 1 \quad & s'_{xo} = s'_{yo} = 0 \\ t_{xo} = t_{yo} = 0 \quad & t'_{xo} = t'_{yo} = 1/\sqrt{\Phi_o} \end{aligned} \tag{3.5}$$

$$\begin{aligned} g_{xo} = g_{yo} = 1 \quad & g_{xa} = g_{ya} = 0 \\ h_{xo} = h_{yo} = 0 \quad & h_{xa} = h_{ya} = 1 \end{aligned} \tag{3.6}$$

$$\begin{aligned} \sigma'_{xo} = \sigma'_{yo} = 1/\sqrt{\Phi_o} \quad & \sigma'_{xa} = \sigma'_{ya} = 0 \\ \tau'_{xo} = \tau'_{yo} = 0 \quad & \tau'_{xa} = \tau'_{ya} = 1/\sqrt{\Phi_a}\,. \end{aligned} \tag{3.7}$$

In the last situation it seems to be more natural to write

$$\begin{aligned} p^{(1)}(z) &= p_o\, \sigma_p(z) + p_a\, \tau_p(z) \\ q^{(1)}(z) &= q_o\, \sigma_q(z) + q_a\, \tau_q(z)\,, \end{aligned} \tag{3.8}$$

so that σ satisfies the same boundary conditions as g and τ as h. Just as p and q were derived from x and y by differentiation ($p = \sqrt{\Phi}\, x'$, $q = \sqrt{\Phi}\, y'$), so are p and q now differentiated to give x and y:

$$x = -\frac{4\sqrt{\Phi}}{\Phi'' - D + 4\eta Q \sqrt{\Phi}}\, p' \quad y = -\frac{4\sqrt{\Phi}}{\Phi'' + D - 4\eta Q \sqrt{\Phi}}\, q'\,. \tag{3.9}$$

(In free space, these expressions become indeterminate, but this is no limitation.)

In general, therefore, we have

$$\begin{aligned} x^{(3)\prime\prime} + \frac{1}{2}\frac{\Phi'}{\Phi} x^{(3)\prime} + \frac{\Phi'' - D + 4\eta Q\sqrt{\Phi}}{4\Phi} x^{(3)} &= X \\ y^{(3)\prime\prime} + \frac{1}{2}\frac{\Phi'}{\Phi} y^{(3)\prime} + \frac{\Phi'' + D - 4\eta Q\sqrt{\Phi}}{4\Phi} y^{(3)} &= Y \end{aligned} \tag{3.10}$$

in which X and Y denote the expressions obtained when $x = x^{(1)}$, $y = y^{(1)}$ have been substituted into the right-hand sides of the differential equations for x and y. If the aberrations are to be expressed in terms of x_o and p_o, we find

$$
\begin{aligned}
x^{(3)}(z_c) &= t_{xc}\int_o^c s_x X\sqrt{\Phi}\,\mathrm{d}z - s_{xc}\int_o^c t_x X\sqrt{\Phi}\,\mathrm{d}z \\
y^{(3)}(z_c) &= t_{yc}\int_o^c s_y Y\sqrt{\Phi}\,\mathrm{d}z - s_{yc}\int_o^c t_y Y\sqrt{\Phi}\,\mathrm{d}z \\
p^{(3)}(z_c) &= \sqrt{\Phi_c}\,t'_{xc}\int_o^c s_x X\sqrt{\Phi}\,\mathrm{d}z - \sqrt{\Phi_c}\,s'_{xc}\int_o^c t_x X\sqrt{\Phi}\,\mathrm{d}z \\
q^{(3)}(z_c) &= \sqrt{\Phi_c}\,t'_{yc}\int_o^c s_y Y\sqrt{\Phi}\,\mathrm{d}z - \sqrt{\Phi_c}\,s'_{yc}\int_o^c t_y Y\sqrt{\Phi}\,\mathrm{d}z\,.
\end{aligned}
\tag{3.11}
$$

If they are to be expressed in terms of the object and aperture co-ordinates,

$$
\begin{aligned}
k_x x^{(3)}(z_c) &= h_{xc}\int_a^c g_x X\sqrt{\Phi}\,\mathrm{d}z - g_{xc}\int_o^c h_x X\sqrt{\Phi}\,\mathrm{d}z \\
k_y y^{(3)}(z_c) &= h_{yc}\int_a^c g_y Y\sqrt{\Phi}\,\mathrm{d}z - g_{yc}\int_o^c h_y Y\sqrt{\Phi}\,\mathrm{d}z
\end{aligned}
\tag{3.12}
$$

in which the constants k_x and k_y denote $\sqrt{\Phi}\,(g_x h'_x - g'_x h_x)$ and $\sqrt{\Phi}\,(g_y h'_y - g'_y h_y)$ respectively.

When the coordinates p_o, q_o, p_a and q_a are employed, similar results can be obtained using σ_x, τ_x, σ_y and τ_y, but it is more natural to work directly in terms of the differential equations for p and q. We normally write

$$p' - x/\xi(z) = 0 \quad q' - y/\eta(z) = 0 \tag{3.13}$$

and to obtain differential equations for x and y, we substitute $p = \partial m^{(2)}/\partial x' = \sqrt{\Phi}\,x'$, $q = \partial m^{(2)}/\partial y' = \sqrt{\Phi}\,y'$; alternatively, however, we may use equations (3.13) to eliminate x and y, thus deriving differential equations for p and q:

$$p = \sqrt{\Phi}\,\mathrm{d}(\xi p')/\mathrm{d}z\,, \quad q = \sqrt{\Phi}\,\mathrm{d}(\eta q')/\mathrm{d}z$$

or

$$p'' + \frac{\xi'}{\xi}p' - \frac{p}{\sqrt{\Phi}\,\xi} = 0\,, \quad q'' + \frac{\eta'}{\eta}q' - \frac{q}{\sqrt{\Phi}\,\eta} = 0\,. \tag{3.14}$$

The functions $\sigma_p(z)$ and $\tau_p(z)$ are linearly independent solutions of the p-equation and $\sigma_q(z)$, $\tau_q(z)$, of the q-equation. The aberrations $p^{(3)}$ and $q^{(3)}$ now satisfy the differential equations

$$
\begin{aligned}
p^{(3)''} + \frac{\xi'}{\xi}p^{(3)'} - \frac{p^{(3)}}{\sqrt{\Phi}\,\xi} &= P \\
q^{(3)''} + \frac{\eta'}{\eta}q^{(3)'} - \frac{q^{(3)}}{\sqrt{\Phi}\,\eta} &= Q
\end{aligned}
\tag{3.15}
$$

and hence

$$p_c^{(3)} = \frac{\tau_{pc}}{k_p} \int\limits_a^c \sigma_p \, P \, \xi \, \mathrm{d}z - \frac{\sigma_{pc}}{k_p} \int\limits_o^c \tau_p \, P \, \xi \, \mathrm{d}z$$
$$q_c^{(3)} = \frac{\tau_{qc}}{k_q} \int\limits_a^c \sigma_q \, Q \, \eta \, \mathrm{d}z - \frac{\sigma_{qc}}{k_q} \int\limits_o^c \tau_q \, Q \, \eta \, \mathrm{d}z \qquad (3.16)$$

in which $k_p = \xi(\sigma_p \tau_p' - \sigma_p' \tau_p)$ and $k_q = \eta(\sigma_q \tau_q' - \sigma_q' \tau_q)$ are again constants.

In the past, the paraxial solutions have occasionally been written

$$x = x_o \, x_\gamma(z) + \frac{x_\alpha(z)}{x_\alpha(z_a)} x_a$$
$$y = y_o \, y_\gamma(z) + \frac{y_\alpha(z)}{y_\alpha(z_a)} y_a \qquad (3.17)$$

in which the solutions of the paraxial equations, x_γ, x_α, y_γ and y_α, satisfy the boundary conditions

$$x_{\gamma o} = y_{\gamma o} = 1 \qquad x_{\gamma a} = y_{\gamma a} = 0$$
$$x_{\alpha o} = y_{\alpha o} = 0 \qquad x'_{\alpha o} = y'_{\alpha o} = 1\,. \qquad (3.18)$$

These are not satisfactory, however, and the formulae for the aberration coefficients which have been deduced with their aid may, when some plane other than a stigmatic image plane is in question, be applicable only when the aperture lies in the field-free region between the object and the lenses. (This point is elaborated in [93]. For a comparison of the correct formulae with those applicable only to the special case mentioned, see my analysis of cylindrical lenses [94].)

If one of the sets of expressions for $x^{(1)}$ and $y^{(1)}$ is substituted into X and Y, the aberration coefficients which result are extremely complicated in appearance. They can be simplified by partial integration and use of the paraxial equations, but this is an exceedingly laborious procedure, rendered still more unattractive by the fact that at the outset, the kind of expression that each integral can eventually be transformed into is far from obvious. It is one of the virtues of aberration theory using perturbation characteristics that all possible forms of the aberration coefficients can be deduced with little extra effort. For this reason, we shall not consider the trajectory method further; expressions for all the aberration coefficients, obtained by substituting $x^{(1)} = g_x x_o + h_x x_a$, $y^{(1)} = g_y y_o + h_y y_a$ into X and Y, are given by *Melkich* [133], and some idea of the labour involved in recasting these can be gained from the appendix to *Hawkes* [89]. (Although *Melkich* uses the rays $x_\alpha \ldots y_\gamma$, his results are not invalidated since the limits to the integrals which appear in his formulae for the aberrations ([133], p. 427, equation 19) are those appropriate to $g_x \ldots h_y$; his $x_\alpha/x_{\alpha B}, \ldots$ may be regarded as $h_x, \ldots$.)

The formulae derived by *Meads* [130] represent the aberration coefficients of magnetic quadrupoles only, and we therefore consider next the method of perturbation characteristic functions, and use it to derive expressions for all the aberration coefficients of twist-free systems.

3.2 Perturbation characteristics

This method was first applied in electron optics by *Glaser* [71] and was subsequently perfected by *Sturrock* [214, 215]. For a complete account, *Sturrock* [214, 217] should be consulted, and additional explanatory material may be found in *Hawkes* [96].

The point characteristic, V, is defined by

$$V_{\alpha\beta} = \int_{\alpha}^{\beta} m \, dz$$

and if m is altered slightly to $m + m^I$, we find that

$$\delta V^I_{\alpha\beta} = (p^I_\beta \, \delta x_\beta + q^I_\beta \, \delta y_\beta - x^I_\beta \, \delta p_\beta - y^I_\beta \, \delta q_\beta) - \\ - (p^I_\alpha \, \delta x_\alpha + q^I_\alpha \, \delta y_\alpha - x^I_\alpha \, \delta p_\alpha - y^I_\alpha \, \delta q_\alpha)$$

in which

$$V^I_{\alpha\beta} = \int_{\alpha}^{\beta} m^I \, dz \, . \tag{3.19}$$

We are at liberty to set any four of the coordinate variations equal to zero, subject to various conditions restricting the type of imagery that the system produces, and, of the various possibilities, we shall be concerned with the point perturbation characteristic, $V^I(x_\alpha, y_\alpha, x_\beta, y_\beta)$, the angle perturbation characteristic $T^I(p_\alpha, q_\alpha, p_\beta, q_\beta)$ and the object perturbation characteristic $U^I(x_\beta, y_\beta, p_\beta, q_\beta)$; we have

$$\delta V^I_{\alpha\beta} = p^I_\beta \, \delta x_\beta + q^I_\beta \, \delta y_\beta - (p^I_\alpha \, \delta x_\alpha + q^I_\alpha \, \delta y_\alpha) \\ (x^I_\alpha = y^I_\alpha = x^I_\beta = y^I_\beta = 0) \, ; \tag{3.20a}$$

$$\delta T^I_{\alpha\beta} = -(x^I_\beta \, \delta p_\beta + y^I_\beta \, \delta q_\beta) + x^I_\alpha \, \delta p_\alpha + y^I_\alpha \, \delta q_\alpha \\ (p^I_\alpha = q^I_\alpha = p^I_\beta = q^I_\beta = 0) \, ; \tag{3.20b}$$

$$\delta U^I_{\alpha\beta} = p^I_\beta \, \delta x_\beta + q^I_\beta \, \delta y_\beta - (x^I_\beta \, \delta p_\beta + y^I_\beta \, \delta q_\beta) \\ (p^I_\alpha = q^I_\alpha = x^I_\alpha = y^I_\alpha = 0) \, . \tag{3.20c}$$

To obtain the aberrations in a system with an aperture, we consider V^I_{oc} and V^I_{ac}. We have

$$\delta V^I_{oc} = p^I_c \, \delta x_c + q^I_c \, \delta y_c - x^I_c \, \delta p_c - y^I_c \, \delta q_c - p^I_o \, \delta x_o - q^I_o \, \delta y_o$$

$$\delta V^I_{ac} = p^I_c \, \delta x_c + q^I_c \, \delta y_c - x^I_c \, \delta p_c - y^I_c \, \delta q_c - p^I_a \, \delta x_a - q^I_a \, \delta y_a$$

and writing $x_c = x_c^{(1)} = x_o g_{xc} + x_a h_{xc}$, $y_c = y_c^{(1)} = y_o g_{yc} + y_a h_{yc}$, we obtain

$$\begin{aligned} x_c^{(3)} = x^I &= \frac{h_{xc}}{k_x}\frac{\partial V_{ac}^{(4)}}{\partial x_o} - \frac{g_{xc}}{k_x}\frac{\partial V_{oc}^{(4)}}{\partial x_a} \\ y_c^{(3)} = y^I &= \frac{h_{yc}}{k_y}\frac{\partial V_{ac}^{(4)}}{\partial y_o} - \frac{g_{yc}}{k_y}\frac{\partial V_{oc}^{(4)}}{\partial y_a} \end{aligned} \tag{3.21}$$

for the primary geometrical aberrations for which $m^I = m^{(4)}$ and $V^{(4)} = \int m^{(4)}\,\mathrm{d}z$. Alternatively, we may use $T^{(4)} = \int m^{(4)}\,\mathrm{d}z$ with $p_c = p_c^{(1)} = p_o \sigma_{pc} + p_a \tau_{pc}$, $q_c = q_c^{(1)} = q_o \sigma_{qc} + q_a \tau_{qc}$, to give

$$\begin{aligned} p_c^{(3)} = p^I &= -\frac{\tau_{pc}}{k_p}\frac{\partial T_{ac}^{(4)}}{\partial p_o} + \frac{\sigma_{pc}}{k_p}\frac{\partial T_{oc}^{(4)}}{\partial p_a} \\ q_c^{(3)} = q^I &= -\frac{\tau_{qc}}{k_q}\frac{\partial T_{ac}^{(4)}}{\partial q_o} + \frac{\sigma_{qc}}{k_q}\frac{\partial T_{oc}^{(4)}}{\partial q_a}\,. \end{aligned} \tag{3.22}$$

If the aperture coordinates are to be supplanted by p_o and q_o, we use $U_{oc}^{(4)}$,

$$\delta U_{oc}^{(4)} = p_c^{(3)}\,\delta x_c + q_c^{(3)}\,\delta y_c - (x_c^{(3)}\,\delta p_c + y_c^{(3)}\,\delta q_c)$$

giving

$$\begin{aligned} x_c^{(3)} &= t_{xc}\frac{\partial U_{oc}^{(4)}}{\partial x_o} - s_{xc}\frac{\partial U_{oc}^{(4)}}{\partial p_o} \\ y_c^{(3)} &= t_{yc}\frac{\partial U_{oc}^{(4)}}{\partial y_o} - s_{yc}\frac{\partial U_{oc}^{(4)}}{\partial q_o} \end{aligned} \tag{3.23a}$$

and

$$\begin{aligned} p_c^{(3)} &= \sqrt{\Phi_c}\left(t'_{xc}\frac{\partial U_{oc}^{(4)}}{\partial x_o} - s'_{xc}\frac{\partial U_{oc}^{(4)}}{\partial p_o}\right) \\ q_c^{(3)} &= \sqrt{\Phi_c}\left(t'_{yc}\frac{\partial U_{oc}^{(4)}}{\partial y_o} - s'_{yc}\frac{\partial U_{oc}^{(4)}}{\partial q_o}\right). \end{aligned} \tag{3.23b}$$

The transformations which lead from the point characteristic function to the angle and two mixed characteristics are Legendre transformations, and it may seem surprising, therefore, that the various perturbation characteristic functions have been obtained so simply, as special cases of the general expression for δV^I. This is a peculiarity of the first-order perturbation theory of rectilinear systems, as *Focke* has shown ([65] and [66], equation 2.30); the relationships are more complex when higher order aberrations are under consideration.

The aberration coefficients will appear as integrals, the integrands of which will be determined by the form of the expression for $\int m^{(4)}\,\mathrm{d}z$. If, therefore, the various possible forms of the latter can be regarded as special cases of a more general expression, in which all the possible partial integrations are taken into account, the task of transforming the aberration coefficients into convenient forms by partial integration will become appreciably less arduous. A procedure for doing this in the case of round lenses has been evolved by *Seman* [167–170] and is extended to quadrupole lenses in *Hawkes* [95, 97]. To take full advantage of the symmetry of the system, we introduce the vectors $\boldsymbol{u} = (x, y)$ and $\boldsymbol{u}^* = (x, -y)$ so that

$$m^{(2)} = -\frac{\Phi''}{8\sqrt{\Phi}}\boldsymbol{u}^2 + \frac{D - 4\eta Q\sqrt{\Phi}}{8\sqrt{\Phi}}\boldsymbol{u}\cdot\boldsymbol{u}^* + \frac{1}{2}\sqrt{\Phi}\,\boldsymbol{u}'^2$$

$$\begin{aligned} m^{(4)} = {} & \left(\frac{\Phi^{(\mathrm{iv})}}{128\sqrt{\Phi}} - \frac{\Phi''^2}{128\Phi^{3/2}} - \frac{1}{2}\mathfrak{D}_1\right)\boldsymbol{u}^4 + \\ & + \left(-\frac{D^2}{128\Phi^{3/2}} + \mathfrak{D}_1\right)(\boldsymbol{u}\cdot\boldsymbol{u}^*)^2 + \\ & + \left(\frac{D\Phi''}{64\Phi^{3/2}} - \frac{D''}{96\sqrt{\Phi}} + \frac{\eta}{48}Q''\right)\boldsymbol{u}^2(\boldsymbol{u}\cdot\boldsymbol{u}^*) - \\ & - \frac{\Phi''}{16\sqrt{\Phi}}\boldsymbol{u}^2\boldsymbol{u}'^2 + \frac{D}{16\sqrt{\Phi}}(\boldsymbol{u}\cdot\boldsymbol{u}^*)\,\boldsymbol{u}'^2 - \frac{1}{8}\sqrt{\Phi}\,\boldsymbol{u}'^4 + \\ & + \frac{\eta}{8}Q'(\boldsymbol{u}^*\times\boldsymbol{u})(\boldsymbol{u}\times\boldsymbol{u}') \end{aligned} \tag{3.24}$$

in which $(a, b) \times (c, d)$ stands for the scalar quantity $ad - bc$; $\mathfrak{D}_1$ denotes $D_1/\sqrt{\Phi} - 2\eta\, Q_1$. The terms of which $m^{(2)}$ and $m^{(4)}$ are composed can be analysed "dimensionally", according to their dependence upon the potential functions, the position vectors and the differential operator d/dz which we denote by $\mathfrak{d}$. The dimensions of the terms of $m^{(2)}$ are thus

$$[\sqrt{\Phi}]^1\,[\boldsymbol{u}^2]^1\,[\mathfrak{d}]^2 \quad \text{and} \quad [D/\sqrt{\Phi} \text{ or } Q]^1\,[\boldsymbol{u}\cdot\boldsymbol{u}^*]^1$$

and of $m^{(4)}$,

$$[\sqrt{\Phi}]^1\,[\boldsymbol{u}^2]^2\,[\mathfrak{d}]^4, \quad [D/\sqrt{\Phi} \text{ or } Q]^1\,[\boldsymbol{u}^2]^1\,[\boldsymbol{u}\cdot\boldsymbol{u}^*]^1\,[\mathfrak{d}]^2$$
$$\text{and} \quad [\sqrt{\Phi}]^{-1}\,[D/\sqrt{\Phi} \text{ or } Q]^2\,[\boldsymbol{u}\cdot\boldsymbol{u}^*]^2$$

together with

$$[\mathfrak{D}_1]^1\,[\boldsymbol{u}^2]^2 \quad \text{and} \quad [\mathfrak{D}_1]^1\,[\boldsymbol{u}\cdot\boldsymbol{u}^*]^2\,.$$

The dimensions of the terms of $\int m^{(4)}\,\mathrm{d}z$ differ from those of $m^{(4)}$ by a factor $[\mathfrak{d}]^{-1}$, so that only $[\sqrt{\Phi}]^1\,[\boldsymbol{u}^2]^2\,[\mathfrak{d}]^3$ and $[D/\sqrt{\Phi}$ or $Q]^1\,[\boldsymbol{u}^2]^1\,[\boldsymbol{u}\cdot\boldsymbol{u}^*]^1 \times$ $\times\,[\mathfrak{d}]^1$ still contain $\mathfrak{d}$. These two expressions generate the following sixteen terms, all with the desired dimensions:

$$\begin{array}{lll} r_1 = \sqrt{\Phi}\,\boldsymbol{u}^{2\prime}\,\boldsymbol{u}'^2 & r_6 = \dfrac{\Phi'''}{\sqrt{\Phi}}\boldsymbol{u}^4 & e_1 = \dfrac{D}{\sqrt{\Phi}}\boldsymbol{u}^{2\prime}(\boldsymbol{u}\cdot\boldsymbol{u}^*) \\ r_2 = \dfrac{\Phi'}{\sqrt{\Phi}}(\boldsymbol{u}^{2\prime})^2 & r_7 = \dfrac{\Phi''\Phi'}{\Phi^{3/2}}\boldsymbol{u}^4 & e_2 = \dfrac{D}{\sqrt{\Phi}}\boldsymbol{u}^2(\boldsymbol{u}\cdot\boldsymbol{u}^*)' \\ r_3 = \dfrac{\Phi'}{\sqrt{\Phi}}\boldsymbol{u}^2\boldsymbol{u}'^2 & r_8 = \dfrac{\Phi'^3}{\Phi^{5/2}}\boldsymbol{u}^4 & e_3 = \dfrac{D'}{\sqrt{\Phi}}\boldsymbol{u}^2(\boldsymbol{u}\cdot\boldsymbol{u}^*) \\ r_4 = \dfrac{\Phi''}{\sqrt{\Phi}}\boldsymbol{u}^2\boldsymbol{u}^{2\prime} & r_9 = \dfrac{D\Phi'}{\Phi^{3/2}}\boldsymbol{u}^2(\boldsymbol{u}\cdot\boldsymbol{u}^*) & m_1 = \eta\, Q\,\boldsymbol{u}^{2\prime}(\boldsymbol{u}\cdot\boldsymbol{u}^*) \\ r_5 = \dfrac{\Phi'^2}{\Phi^{3/2}}\boldsymbol{u}^2\boldsymbol{u}^{2\prime} & r_{10} = \eta\dfrac{Q\Phi'}{\Phi}\boldsymbol{u}^2(\boldsymbol{u}\cdot\boldsymbol{u}^*) & m_2 = \eta\, Q\,\boldsymbol{u}^2(\boldsymbol{u}\cdot\boldsymbol{u}^*)' \\ & & m_3 = \eta\, Q'\,\boldsymbol{u}^2(\boldsymbol{u}\cdot\boldsymbol{u}^*)\,. \end{array} \tag{3.25}$$

On differentiating each of the r_i, e_i and m_i, and eliminating $\boldsymbol{u}''$ by means of the paraxial equation of motion,

$$\boldsymbol{u}'' = -\frac{\Phi'}{2\,\Phi}\boldsymbol{u}' - \frac{\Phi''}{4\,\Phi}\boldsymbol{u} + \frac{D - 4\,\eta\,Q\sqrt{\Phi}}{4\,\Phi}\boldsymbol{u}^*\,,$$

we obtain a set of sixteen relations of the form

$$r_i' + \bar{r}_i = 0\,,\quad e_i' + \bar{e}_i = 0\,,\quad m_i' + \overline{m}_i = 0$$

or

$$[r_i]_\alpha^\beta + \int_\alpha^\beta \bar{r}_i\,\mathrm{d}z = 0\,,\quad [e_i]_\alpha^\beta + \int_\alpha^\beta \bar{e}_i\,\mathrm{d}z = 0\,,\quad [m_i]_\alpha^\beta + \int_\alpha^\beta \overline{m}_i\,\mathrm{d}z = 0\,.$$

By adding arbitrary multiples of these equations to $V_{\alpha\beta}^{(4)}$, we obtain the most general expression for the point (or angle or object) characteristic function; any special case can be extracted by choosing the multipliers appropriately. The expressions $\bar{r}_i$, $\bar{e}_i$ and $\overline{m}_i$ involve the following forty terms:

$$R_1 = \frac{\Phi'^4}{\Phi^{7/2}}\boldsymbol{u}^4 \qquad R_9 = \eta\frac{Q\,\Phi''}{\Phi}\boldsymbol{u}^2(\boldsymbol{u}\cdot\boldsymbol{u}^*) \qquad R_{17} = \frac{\Phi''}{\sqrt{\Phi}}\boldsymbol{u}^2\,\boldsymbol{u}'^2$$

$$R_2 = \frac{\Phi'^2\,\Phi''}{\Phi^{5/2}}\boldsymbol{u}^4 \qquad R_{10} = \eta\frac{Q\,\Phi'^2}{\Phi^2}\boldsymbol{u}^2(\boldsymbol{u}\cdot\boldsymbol{u}^*) \qquad R_{18} = \frac{\Phi'^2}{\Phi^{3/2}}\boldsymbol{u}^2\,\boldsymbol{u}'^2$$

$$R_3 = \frac{\Phi''^2}{\Phi^{3/2}}\boldsymbol{u}^4 \qquad R_{11} = \eta\frac{Q'\,\Phi'}{\Phi}\boldsymbol{u}^2(\boldsymbol{u}\cdot\boldsymbol{u}^*) \qquad R_{19} = \eta\frac{Q\,\Phi'}{\Phi}\boldsymbol{u}^{2\prime}(\boldsymbol{u}\cdot\boldsymbol{u}^*)$$

$$R_4 = \frac{\Phi'''\,\Phi'}{\Phi^{3/2}}\boldsymbol{u}^4 \qquad R_{12} = \frac{\Phi'^3}{\Phi^{5/2}}\boldsymbol{u}^2\,\boldsymbol{u}^{2\prime} \qquad R_{20} = \eta\frac{Q\,\Phi'}{\Phi}\boldsymbol{u}^2(\boldsymbol{u}\cdot\boldsymbol{u}^*)'$$

$$R_5 = \frac{\Phi^{(\mathrm{iv})}}{\sqrt{\Phi}}\boldsymbol{u}^4 \qquad R_{13} = \frac{\Phi''\,\Phi'}{\Phi^{3/2}}\boldsymbol{u}^2\,\boldsymbol{u}^{2\prime} \qquad R_{21} = \frac{\Phi'^2}{\Phi^{3/2}}(\boldsymbol{u}^{2\prime})^2$$

$$R_6 = \frac{D\,\Phi''}{\Phi^{3/2}}\boldsymbol{u}^2(\boldsymbol{u}\cdot\boldsymbol{u}^*) \qquad R_{14} = \frac{\Phi'''}{\sqrt{\Phi}}\boldsymbol{u}^2\,\boldsymbol{u}^{2\prime} \qquad R_{22} = \frac{\Phi''}{\sqrt{\Phi}}(\boldsymbol{u}^{2\prime})^2$$

$$R_7 = \frac{D\,\Phi'^2}{\Phi^{5/2}}\boldsymbol{u}^2(\boldsymbol{u}\cdot\boldsymbol{u}^*) \qquad R_{15} = \frac{D\,\Phi'}{\Phi^{3/2}}\boldsymbol{u}^{2\prime}(\boldsymbol{u}\cdot\boldsymbol{u}^*) \qquad R_{23} = \frac{\Phi'}{\sqrt{\Phi}}\boldsymbol{u}^{2\prime}\,\boldsymbol{u}'^2$$

$$R_8 = \frac{D'\,\Phi'}{\Phi^{3/2}}\boldsymbol{u}^2(\boldsymbol{u}\cdot\boldsymbol{u}^*) \qquad R_{16} = \frac{D\,\Phi'}{\Phi^{3/2}}\boldsymbol{u}^2(\boldsymbol{u}\cdot\boldsymbol{u}^*)' \qquad R_{24} = \sqrt{\Phi}\,\boldsymbol{u}'^4$$

$$E_1 = \frac{D'}{\sqrt{\Phi}}\boldsymbol{u}^{2\prime}(\boldsymbol{u}\cdot\boldsymbol{u}^*) \qquad M_1 = \eta\,Q'\,\boldsymbol{u}^{2\prime}(\boldsymbol{u}\cdot\boldsymbol{u}^*)$$

$$E_2 = \frac{D'}{\sqrt{\Phi}}\boldsymbol{u}^2(\boldsymbol{u}\cdot\boldsymbol{u}^*)' \qquad M_2 = \eta\,Q'\,\boldsymbol{u}^2(\boldsymbol{u}\cdot\boldsymbol{u}^*)'$$

$$E_3 = \frac{D^2}{\Phi^{3/2}}(\boldsymbol{u}\cdot\boldsymbol{u}^*)^2 \qquad M_3 = \eta^2\frac{Q^2}{\sqrt{\Phi}}(\boldsymbol{u}\cdot\boldsymbol{u}^*)^2$$

$$E_4 = \frac{D^2}{\Phi^{3/2}}\boldsymbol{u}^4 \qquad M_4 = \eta^2\frac{Q^2}{\sqrt{\Phi}}\boldsymbol{u}^4$$

$$E_5 = \frac{D}{\sqrt{\Phi}}\boldsymbol{u}'^2(\boldsymbol{u}\cdot\boldsymbol{u}^*) \qquad M_5 = \eta\,Q\,\boldsymbol{u}'^2(\boldsymbol{u}\cdot\boldsymbol{u}^*)$$

$$E_6 = \frac{D}{\sqrt{\Phi}}\boldsymbol{u}^{2\prime}(\boldsymbol{u}\cdot\boldsymbol{u}^*)' \qquad M_6 = \eta\,Q\,\boldsymbol{u}^{2\prime}(\boldsymbol{u}\cdot\boldsymbol{u}^*)'$$

$$E_7 = \frac{D''}{\sqrt{\Phi}}\boldsymbol{u}^2(\boldsymbol{u}\cdot\boldsymbol{u}^*) \qquad M_7 = \eta\,Q''\,\boldsymbol{u}^2(\boldsymbol{u}\cdot\boldsymbol{u}^*)$$

$$E_8 = \eta\frac{D\,Q}{\Phi}(\boldsymbol{u}\cdot\boldsymbol{u}^*)^2 \qquad M_8 = \eta\frac{D\,Q}{\Phi}\boldsymbol{u}^4\,. \qquad (3.26)$$

In terms of these quantities, we find:

$$r_1' + {}^1/_2\, R_{17} + {}^1/_4\, R_{22} + R_{23} - 2R_{24} - {}^1/_2\, E_5 - {}^1/_4\, E_6 + 2M_5 + M_6 = 0$$
$$r_2' + R_{13} - R_{15} + 4R_{19} + {}^3/_2\, R_{21} - R_{22} - 4R_{23} = 0$$
$$r_3' + {}^1/_4\, R_{13} - {}^1/_4\, R_{16} - R_{17} + {}^3/_2\, R_{18} + R_{20} - R_{23} = 0$$
$$r_4' + {}^1/_2\, R_3 - {}^1/_2\, R_6 + 2R_9 + R_{13} - R_{14} - 2R_{17} - R_{22} = 0$$
$$r_5' + {}^1/_2\, R_2 - {}^1/_2\, R_7 + 2R_{10} + 2R_{12} - 2R_{13} - 2R_{18} - R_{21} = 0$$
$$r_6' + {}^1/_2\, R_4 - R_5 - 2R_{14} = 0$$
$$r_7' + {}^3/_2\, R_2 - R_3 - R_4 - 2R_{13} = 0$$
$$r_8' + {}^5/_2\, R_1 - 3R_2 - 2R_{12} = 0$$
$$r_9' - R_6 + {}^3/_2\, R_7 - R_8 - R_{15} - R_{16} = 0$$
$$r_{10}' - R_9 + R_{10} - R_{11} - R_{19} - R_{20} = 0$$
$$e_1' + {}^1/_2\, R_6 + R_{15} - E_1 - {}^1/_2\, E_3 - 2E_5 - E_6 + 2E_8 = 0$$
$$e_2' + {}^1/_2\, R_6 + R_{16} - E_2 - {}^1/_2\, E_4 + 2E_5 - 2E_6 + 2M_8 = 0$$
$$e_3' + {}^1/_2\, R_8 - E_1 - E_2 - E_7 = 0$$
$$m_1' + {}^1/_2\, R_9 + {}^1/_2\, R_{19} - {}^1/_2\, E_8 - M_1 + 2M_3 - 2M_5 - M_6 = 0$$
$$m_2' + {}^1/_2\, R_9 + {}^1/_2\, R_{20} - M_2 + 2M_4 + 2M_5 - 2M_6 - {}^1/_2\, M_8 = 0$$
$$m_3' - M_1 - M_2 - M_7 = 0\,.$$

The perturbation characteristic $V^{(4)}$ is given by

$$\begin{aligned} V^{(4)} = \int m^{(4)}\,\mathrm{d}z = {}^1/_2 \int \mathfrak{D}_1\{-\boldsymbol{u}^4 + 2(\boldsymbol{u}\cdot\boldsymbol{u}^*)^2\}\,\mathrm{d}z + \\ + {}^1/_{128} \int (-R_3 + R_5 + 2R_6 - 8R_{17} - 16R_{24} - \\ - E_3 + 8E_5 - {}^4/_3\, E_7 + 8M_1 - 8M_2 + {}^8/_3\, M_7)\,\mathrm{d}z \end{aligned} \tag{3.27}$$

and will be unaltered in value by the addition of $\int \sum_i \{\varrho_i(r_i' + \bar{r}_i) + \varepsilon_i(e_i' + \bar{e}_i) + \mu_i(m_i' + \bar{m}_i)\}\,\mathrm{d}z$, in which ϱ_i, ε_i and μ_i denote arbitrary multipliers. The general form of $V^{(4)}$ is thus

$$128 V^{(4)} - 64 \int \mathfrak{D}_1\{-\boldsymbol{u}^4 + 2(\boldsymbol{u}\cdot\boldsymbol{u}^*)^2\}\,\mathrm{d}z$$
$$= \left[\sum_{i=1}^{10} \varrho_i r_i + \sum_{i=1}^{3} \varepsilon_i e_i + \sum_{i=1}^{3} \mu_i m_i\right] +$$
$$+ \int \{{}^5/_2\, \varrho_8 R_1 + ({}^1/_2\, \varrho_5 + {}^3/_2\, \varrho_7 - 3\varrho_8) R_2 + (-1 + {}^1/_2\, \varrho_4 - \varrho_7) R_3 +$$
$$+ ({}^1/_2\, \varrho_6 - \varrho_7) R_4 + (1 - \varrho_6) R_5 + (2 - {}^1/_2\, \varrho_4 - \varrho_9 + {}^1/_2\, \varepsilon_1 + {}^1/_2\, \varepsilon_2) R_6 +$$
$$+ (-{}^1/_2\, \varrho_5 + {}^3/_2\, \varrho_9) R_7 + (-\varrho_9 + {}^1/_2\, \varepsilon_3) R_8 +$$
$$+ (2\varrho_4 - \varrho_{10} + {}^1/_2\, \mu_1 + {}^1/_2\, \mu_2) R_9 + (2\varrho_5 + \varrho_{10}) R_{10} - \varrho_{10} R_{11} +$$
$$+ 2(\varrho_5 - \varrho_8) R_{12} +$$
$$+ (\varrho_2 + {}^1/_4\, \varrho_3 + \varrho_4 - 2\varrho_5 - 2\varrho_7) R_{13} + (-\varrho_4 - 2\varrho_6) R_{14} +$$

$$+ (-\varrho_2 - \varrho_9 + \varepsilon_1) R_{15} + (-{}^1/_4\, \varrho_3 - \varrho_9 + \varepsilon_2) R_{16} +$$
$$+ (-8 + {}^1/_2\, \varrho_1 - \varrho_3 - 2\varrho_4) R_{17} + ({}^3/_2\, \varrho_3 - 2\varrho_5) R_{18} +$$
$$+ (4\varrho_2 - \varrho_{10} + {}^1/_2\, \mu_1) R_{19} + (\varrho_3 - \varrho_{10} + {}^1/_2\, \mu_2) R_{20} + ({}^3/_2\, \varrho_2 - \varrho_5) R_{21} +$$
$$+ ({}^1/_4\, \varrho_1 - \varrho_2 - \varrho_4) R_{22} + (\varrho_1 - 4\varrho_2 - \varrho_3) R_{23} + (-16 - 2\varrho_1) R_{24} +$$
$$+ (-\varepsilon_1 - \varepsilon_3) E_1 + (-\varepsilon_2 - \varepsilon_3) E_2 + (-1 - {}^1/_2\, \varepsilon_1) E_3 -$$
$$- {}^1/_2\, \varepsilon_2 E_4 + (8 - {}^1/_2\, \varrho_1 - 2\varepsilon_1 + 2\varepsilon_2) E_5 + (-{}^1/_4\, \varrho_1 - \varepsilon_1 - 2\varepsilon_2) E_6 +$$
$$+ (-{}^4/_3 - \varepsilon_3) E_7 + (2\varepsilon_1 - {}^1/_2\, \mu_1) E_8 +$$
$$+ (8 - \mu_1 - \mu_3) M_1 + (-8 - \mu_2 - \mu_3) M_2 + 2\,\mu_1 M_3 +$$
$$+ 2\,\mu_2 M_4 + (2\varrho_1 - 2\,\mu_1 + 2\,\mu_2) M_5 +$$
$$+ (\varrho_1 - \mu_1 - 2\,\mu_2) M_6 + ({}^8/_3 - \mu_3) M_7 +$$
$$+ (2\varepsilon_2 - {}^1/_2\, \mu_2) M_8\} \,\mathrm{d}z\,. \qquad (3.28)$$

To proceed in a perfectly general manner we should now substitute general expressions for x and y into $V^{(4)}$, $x = A_x x_a(z) + B_x x_b(z)$, $y = A_y y_a(z) + B_y y_b(z)$; by identifying A_x with x_o and B_x with x_a, and hence x_a with g_x and x_b with h_x, we should obtain the aberration coefficients in a completely general form for a system with an aperture; by setting A_x equal to x_o and B_x to p_o, and hence x_a to s_x and x_b to t_x we should obtain them for a system without an aperture; the other possibilities can be treated similarly. The desired form of the coefficients would then be extracted by choosing the multipliers ϱ_i, ε_i and μ_i suitably. The number of desirable forms for the coefficients is limited, however, and it is more reasonable to decide which these are from the outset. If the potential functions are such that some analytical expression represents each adequately, it will be advantageous to eliminate products of derivatives of $\boldsymbol{u}$ as far as possible; if, however, the potential functions are available only in the form of a set of computed values, it will be better either to arrange that as little differentiation as possible has to be done, or to select an expression in which the potential functions D and Q do not have to be differentiated at all and Φ not more than twice. How to achieve these ends is analysed in some detail in [95]. Here we merely list the results.

Products of $\boldsymbol{u}'$ can be largely eliminated by writing

$$\begin{array}{lll}
\varrho_1 = -8 & \varrho_6 = 1 & \varepsilon_1 = 14/3 \\
\varrho_2 = -1/2 \text{ or } 2 & \varrho_7 = 0 & \varepsilon_2 = -4/3 \\
\varrho_3 = -6 \text{ or } -16 & \varrho_8 = -2 & \varepsilon_3 = -5/3 \\
\varrho_4 = -2 & \varrho_9 = 8/3 & \mu_1 = -8 \\
\varrho_5 = -2 & \varrho_{10} = -6 & \mu_2 = 0 \\
 & & \mu_3 = 4
\end{array} \qquad (3.29)$$

so that $V^{(4)}$ becomes

$$128\, V^{(4)} = \Big[-8\sqrt{\Phi}\, \boldsymbol{u}^{2\prime}\, \boldsymbol{u}'^{2} + (-1/2; 2)\frac{\Phi'}{\sqrt{\Phi}}(\boldsymbol{u}^{2\prime})^2 -$$

$$-(6; 16)\frac{\Phi'}{\sqrt{\Phi}}\boldsymbol{u}^2\boldsymbol{u}'^2 - 2\frac{\Phi''}{\sqrt{\Phi}}\boldsymbol{u}^2\boldsymbol{u}^{2\prime} - 2\frac{\Phi'^2}{\Phi^{3/2}}\boldsymbol{u}^2\boldsymbol{u}^{2\prime} +$$

$$+\frac{\Phi'''}{\sqrt{\Phi}}\boldsymbol{u}^4 - 2\frac{\Phi'^3}{\Phi^{5/2}}\boldsymbol{u}^4 + \frac{8}{3}\frac{D\Phi'}{\Phi^{3/2}}\boldsymbol{u}^2(\boldsymbol{u}\cdot\boldsymbol{u}^*) -$$

$$-6\eta\frac{Q\Phi'}{\Phi}\boldsymbol{u}^2(\boldsymbol{u}\cdot\boldsymbol{u}^*) + \frac{14}{3}\frac{D}{\sqrt{\Phi}}\boldsymbol{u}^{2\prime}(\boldsymbol{u}\cdot\boldsymbol{u}^*) - \frac{4D}{3\sqrt{\Phi}}\boldsymbol{u}^2(\boldsymbol{u}\cdot\boldsymbol{u}^*)' -$$

$$-\frac{5}{3}\frac{D'}{\sqrt{\Phi}}\boldsymbol{u}^2(\boldsymbol{u}\cdot\boldsymbol{u}^*) - 8\eta\, Q\boldsymbol{u}^{2\prime}(\boldsymbol{u}\cdot\boldsymbol{u}^*) + 4\eta\, Q'\boldsymbol{u}^2(\boldsymbol{u}\cdot\boldsymbol{u}^*)\Big] +$$

$$+\int\Big[\Big(-64\,\mathfrak{D}_1 - 5\frac{\Phi'^4}{\Phi^{7/2}} + 5\frac{\Phi'^2\Phi''}{\Phi^{5/2}} - 2\frac{\Phi''^2}{\Phi^{3/2}} + \frac{\Phi'''\Phi'}{2\,\Phi^{3/2}} + \frac{2}{3}\frac{D^2}{\Phi^{3/2}} -$$

$$-\frac{8}{3}\eta\frac{DQ}{\Phi}\Big)\boldsymbol{u}^4 +$$

$$+\Big(2\frac{D\Phi''}{\Phi^{3/2}} + 5\frac{D\Phi'^2}{\Phi^{5/2}} - \frac{7D\Phi'}{2\,\Phi^{3/2}} - 2\eta\frac{Q\Phi''}{\Phi} - 10\eta\frac{Q\Phi'^2}{\Phi^2} +$$

$$+6\eta\frac{Q'\Phi'}{\Phi}\Big)\boldsymbol{u}^2(\boldsymbol{u}\cdot\boldsymbol{u}^*) + \Big\{(5;0)\frac{D\Phi'}{\Phi^{3/2}} + (0;20)\,\eta\frac{Q\Phi'}{\Phi}\Big\}(\boldsymbol{u}\times\boldsymbol{u}^*)(\boldsymbol{u}\times\boldsymbol{u}') +$$

$$+\Big\{(-2;8)\frac{\Phi''}{\sqrt{\Phi}} - (5;20)\frac{\Phi'^2}{\Phi^{3/2}}\Big\}(\boldsymbol{u}\times\boldsymbol{u}')^2 -$$

$$-6\Big(\frac{D'}{\sqrt{\Phi}} - 4\eta\, Q'\Big)(\boldsymbol{u}\times\boldsymbol{u}^*)(\boldsymbol{u}\times\boldsymbol{u}') +$$

$$+\Big(128\,\mathfrak{D}_1 - \frac{10\,D^2}{3\,\Phi^{3/2}} + \frac{40}{3}\eta\frac{DQ}{\Phi} - 16\eta^2\frac{Q^2}{\sqrt{\Phi}}\Big)(\boldsymbol{u}\cdot\boldsymbol{u}^*)^2 +$$

$$+\frac{1}{3}\Big(\frac{D''}{\sqrt{\Phi}} - 4\eta\, Q''\Big)\boldsymbol{u}^2(\boldsymbol{u}\cdot\boldsymbol{u}^*)\Big]\,\mathrm{d}z\,. \tag{3.30}$$

If derivatives of D and Q higher than the first are to be avoided, and $\boldsymbol{u}'^2$ and higher terms eliminated, we might select

$$\begin{array}{lll}
\varrho_1 = -8 & \varrho_6 = 1 & \varepsilon_1 = 14/3 \\
\varrho_2 = 0 & \varrho_7 = 1/2 & \varepsilon_2 = -4/3 \\
\varrho_3 = -8 & \varrho_8 = -2 & \varepsilon_3 = -4/3 \\
\varrho_4 = -2 & \varrho_9 = 14/3 & \mu_1 = -8 \\
\varrho_5 = -5/2 & \varrho_{10} = -8 & \mu_2 = 0 \\
 & & \mu_3 = 8/3
\end{array} \tag{3.31}$$

so that of the terms in Φ'', only R_2 and R_3 remain. $V^{(4)}$ is now of the form

$$
\begin{aligned}
128\,V^{(4)} &= 64\int \mathfrak{D}_1\{2(\boldsymbol{u}\cdot\boldsymbol{u}^*)^2 - \boldsymbol{u}^4\}\,\mathrm{d}z + \\
&+ \Big[\sum_i (\varrho_i r_i + \varepsilon_i e_i + \mu_i m_i)\Big] + \\
&+ \int\Big[\Big(-5\frac{\Phi'^4}{\Phi^{7/2}} + \frac{11}{2}\frac{\Phi'^2\Phi''}{\Phi^{5/2}} - \frac{5}{2}\frac{\Phi''^2}{\Phi^{3/2}}\Big)\boldsymbol{u}^4 + \\
&+ \Big(\frac{33}{4}\frac{D\Phi'^2}{\Phi^{5/2}} - \frac{16}{3}\frac{D'\Phi'}{\Phi^{3/2}} - 13\eta\frac{Q\Phi'^2}{\Phi^2} + 8\eta\frac{Q'\Phi'}{\Phi}\Big)\boldsymbol{u}^2(\boldsymbol{u}\cdot\boldsymbol{u}^*) - \\
&- \frac{\Phi'^3}{\Phi^{5/2}}\boldsymbol{u}^2\boldsymbol{u}^{2\prime} - 4\frac{D\Phi'}{\Phi^{3/2}}\boldsymbol{u}^2(\boldsymbol{u}\cdot\boldsymbol{u}^*)' - 7\frac{\Phi'^2}{\Phi^{3/2}}\boldsymbol{u}^2\boldsymbol{u}'^2 + \\
&+ 4\eta\frac{Q\Phi'}{\Phi}\boldsymbol{u}^{2\prime}(\boldsymbol{u}\cdot\boldsymbol{u}^*) + \frac{5}{2}\frac{\Phi'^2}{\Phi^{3/2}}(\boldsymbol{u}^{2\prime})^2 - \frac{10}{3}\Big(\frac{D'}{\sqrt{\Phi}} - 4\eta Q'\Big)\boldsymbol{u}^{2\prime}(\boldsymbol{u}\cdot\boldsymbol{u}^*) + \\
&+ \frac{8}{3}\Big(\frac{D'}{\sqrt{\Phi}} - 4\eta Q'\Big)\boldsymbol{u}^2(\boldsymbol{u}\cdot\boldsymbol{u}^*)' - \Big(\frac{10}{3}\frac{D^2}{\Phi^{3/2}} + 16\eta^2\frac{Q^2}{\sqrt{\Phi}}\Big)(\boldsymbol{u}\cdot\boldsymbol{u}^*)^2 + \\
&+ \frac{2}{3}\frac{D^2}{\Phi^{3/2}}\boldsymbol{u}^4 + \frac{8}{3}\eta\frac{DQ}{\Phi}\{5(\boldsymbol{u}\cdot\boldsymbol{u}^*)^2 - \boldsymbol{u}^4\}\Big]\,\mathrm{d}z\,. \qquad (3.32)
\end{aligned}
$$

For systems not containing any round lens component, $\Phi = \text{const}$, derivatives of D and Q may be removed altogether by writing

$$\varepsilon_1 = \varepsilon_2 = -\varepsilon_3 = 4/3\,,\quad \mu_1 = -\frac{1}{2}\mu_2 = 2\,\mu_3 = 16/3 \qquad (3.33)$$

so that

$$
\begin{aligned}
128\,V^{(4)} &= \Big[\varrho_1\sqrt{\Phi}\,\boldsymbol{u}^{2\prime}\boldsymbol{u}'^2 + \frac{4}{3}\frac{D}{\sqrt{\Phi}}\{\boldsymbol{u}^{2\prime}(\boldsymbol{u}\cdot\boldsymbol{u}^*) + \boldsymbol{u}^2(\boldsymbol{u}\cdot\boldsymbol{u}^*)'\} - \\
&- \frac{4}{3}\frac{D'}{\sqrt{\Phi}}\boldsymbol{u}^2(\boldsymbol{u}\cdot\boldsymbol{u}^*) + \frac{16}{3}\eta Q\{\boldsymbol{u}^{2\prime}(\boldsymbol{u}\cdot\boldsymbol{u}^*) - 2\boldsymbol{u}^2(\boldsymbol{u}\cdot\boldsymbol{u}^*)'\} + \\
&+ \frac{8}{3}\eta Q'\boldsymbol{u}^2(\boldsymbol{u}\cdot\boldsymbol{u}^*)\Big] + \\
&+ 64\int\mathfrak{D}_1\{-\boldsymbol{u}^4 + 2(\boldsymbol{u}\cdot\boldsymbol{u}^*)^2\}\,\mathrm{d}z + \qquad (3.34)\\
&+ \int\Big\{-\frac{5}{3}\frac{D^2}{\Phi^{3/2}}(\boldsymbol{u}\cdot\boldsymbol{u}^*)^2 - \frac{2}{3}\frac{D^2}{\Phi^{3/2}}\boldsymbol{u}^4 + \Big(8 - \frac{\varrho_1}{2}\Big)\frac{D}{\sqrt{\Phi}}\boldsymbol{u}'^2(\boldsymbol{u}\cdot\boldsymbol{u}^*) - \\
&- \Big(\frac{\varrho_1}{4} + 4\Big)\frac{D}{\sqrt{\Phi}}\boldsymbol{u}^{2\prime}(\boldsymbol{u}\cdot\boldsymbol{u}^*)' + \frac{32}{3}\eta^2\frac{Q^2}{\sqrt{\Phi}}(\boldsymbol{u}\cdot\boldsymbol{u}^*)^2 - \\
&- \frac{64}{3}\eta^2\frac{Q^2}{\sqrt{\Phi}}\boldsymbol{u}^4 + (2\varrho_1 - 32)\,\eta\,Q\boldsymbol{u}'^2(\boldsymbol{u}\cdot\boldsymbol{u}^*) + \\
&+ (\varrho_1 + 16)\,\eta\,Q\boldsymbol{u}^{2\prime}(\boldsymbol{u}\cdot\boldsymbol{u}^*)' + 8\eta\frac{DQ}{\Phi}\boldsymbol{u}^4 - (16 + 2\varrho_1)\sqrt{\Phi}\,\boldsymbol{u}'^4\Big\}\,\mathrm{d}z\,.
\end{aligned}
$$

The undetermined multiplier ϱ_1 may be used either to remove $\boldsymbol{u}'^4$ ($\varrho_1 = -8$) or to cast the two pairs of terms in $\boldsymbol{u}'^2(\boldsymbol{u}\cdot\boldsymbol{u}^*)$ and $\boldsymbol{u}^{2\prime}(\boldsymbol{u}\cdot\boldsymbol{u}^*)'$ into the form

$$\frac{32}{3}\Big(\frac{D}{\sqrt{\Phi}} - 4\eta\,Q\Big)(\boldsymbol{u}\times\boldsymbol{u}')(\boldsymbol{u}^*\times\boldsymbol{u}')$$

($\varrho_1 = -16/3$).

In practice, we are rarely concerned with aberrations other than those which do not vanish when the object point is situated on the axis, the "aperture aberrations"; these have the most deleterious effect when the system is to play the role of an objective lens. If, however, the lens system is to be used as a projective, the aberrations which depend only on the position of the object point, the distortions, will dominate, and so the formulae for these are also of interest. We shall, however, see that the formulae for aperture aberrations and distortions are in a sense complementary, and once one set has been calculated, the other can be written down immediately (cf. [90], p. 367). For this reason, only one set of expressions for all the aberration coefficients will be given, from which products of $\boldsymbol{u}'$ have been eliminated as far as possible (ϱ_i, ε_i, μ_i given by equation 3.29); the aperture aberration coefficients (and, by implication, the distortion coefficients) will be listed in a number of other forms. When there is an aperture, we have

$$V^{(4)} = \sum_{\alpha,\beta,\gamma,\delta \geqq 0} (\alpha\,\beta\,\gamma\,\delta)\, x_o^\alpha y_o^\beta x_a^\gamma y_a^\delta \begin{cases} \alpha+\beta+\gamma+\delta = 4 \\ \alpha+\gamma,\ \beta+\delta \quad \text{always even} \end{cases}$$

and in terms of $\boldsymbol{u}$ and $\boldsymbol{u}^*$, we write

$$\begin{aligned} V^{(4)} = {} & [t_1 \boldsymbol{u}^4 + t_2 \boldsymbol{u}^2(\boldsymbol{u}\cdot\boldsymbol{u}^*) + t_3 \boldsymbol{u}^{2\prime} \boldsymbol{u}'^2 + t_4(\boldsymbol{u}^{2\prime})^2 + \\ & + t_5 \boldsymbol{u}^2 \boldsymbol{u}'^2 + t_6 \boldsymbol{u}^2 \boldsymbol{u}^{2\prime} + t_7 \boldsymbol{u}^{2\prime}(\boldsymbol{u}\cdot\boldsymbol{u}^*) + t_8 \boldsymbol{u}^2(\boldsymbol{u}\cdot\boldsymbol{u}^*)'] + \\ & + \int \{T_1 \boldsymbol{u}^4 + T_2 \boldsymbol{u}^2(\boldsymbol{u}\cdot\boldsymbol{u}^*) + T_3(\boldsymbol{u}\cdot\boldsymbol{u}^*)^2 + T_4(\boldsymbol{u}\times\boldsymbol{u}^*)(\boldsymbol{u}\times\boldsymbol{u}') + \\ & + T_5(\boldsymbol{u}\times\boldsymbol{u}')^2\}\,\mathrm{d}z \end{aligned} \tag{3.35}$$

in which

$$128 t_1 = \frac{\Phi'''}{\sqrt{\Phi}} - 2\frac{\Phi'^3}{\Phi^{5/2}};$$

$$128 t_2 = \frac{8}{3}\frac{D\Phi'}{\Phi^{3/2}} - 6\eta Q\frac{\Phi'}{\Phi} - \frac{5}{3}\frac{D'}{\sqrt{\Phi}} + 4\eta Q';$$

$$16 t_3 = -\sqrt{\Phi}; \quad 128 t_4 = (-1/2 \ \text{ or } \ 2)\frac{\Phi'}{\sqrt{\Phi}};$$

$$64 t_5 = -(3 \ \text{ or } \ 8)\frac{\Phi''}{\sqrt{\Phi}}; \quad 64 t_6 = -\frac{\Phi''}{\sqrt{\Phi}} - \frac{\Phi'^2}{\Phi^{3/2}};$$

$$64 t_7 = \frac{7}{3}\frac{D}{\sqrt{\Phi}} - 4\eta Q; \quad 96 t_8 = -\frac{D}{\sqrt{\Phi}};$$

and

$$\begin{aligned} 128 T_1 = {} & -64\mathfrak{D}_1 - 5\frac{\Phi'^4}{\Phi^{7/2}} + 5\frac{\Phi'^2\Phi''}{\Phi^{5/2}} - 2\frac{\Phi''^2}{\Phi^{3/2}}\frac{\Phi'''\Phi'}{2\Phi^{3/2}} + \\ & + \frac{2}{3}\frac{D^2}{\Phi^{3/2}} - \frac{8}{3}\eta\frac{DQ}{\Phi}; \end{aligned}$$

$$\begin{aligned} 128 T_2 = {} & 2\frac{D\Phi''}{\Phi^{3/2}} + 5\frac{D\Phi'^2}{\Phi^{5/2}} - \frac{7}{2}\frac{D\Phi'}{\Phi^{3/2}} - 2\eta\frac{Q\Phi''}{\Phi} - 10\eta\frac{Q\Phi'^2}{\Phi} + \\ & + 6\eta\frac{Q'\Phi'}{\Phi} + \frac{D''}{3\sqrt{\Phi}} - \frac{4}{3}\eta Q''; \end{aligned}$$

$$64 T_3 = 64\mathfrak{D}_1 - \frac{5}{3}\frac{D^2}{\Phi^{3/2}} + \frac{20}{3}\eta\frac{DQ}{\Phi} - 8\eta^2\frac{Q^2}{\sqrt{\Phi}};$$

$$128 T_4 = (5 \ \text{ or } \ 0)\frac{D\Phi'}{\Phi^{3/2}} + (0 \ \text{ or } \ 20)\,\eta\frac{Q\Phi'}{\Phi} - 6\left(\frac{D'}{\sqrt{\Phi}} - 4\eta Q'\right);$$

$$128 T_5 = (-2 \ \text{ or } \ 8)\frac{\Phi''}{\sqrt{\Phi}} - (5 \ \text{ or } \ 20)\frac{\Phi'^2}{\Phi^{3/2}}. \tag{3.36}$$

The aberrations are hence of the form

$$x^{(3)} = \sum_{p,q,r,s \geqq 0} (p\,q\,r\,s)\, x_o^p\, y_o^q\, x_a^r\, y_a^s \begin{cases} p+q+r+s=3 \\ p+r \text{ odd} \end{cases}$$

$$y^{(3)} = \sum_{p,q,r,s, \geqq 0} (p\,q\,r\,s)\, x_o^p\, y_o^q\, x_a^r\, y_a^s \begin{cases} p+q+r+s=3 \\ q+s \text{ odd}\,. \end{cases} \qquad (3.37)$$

The formulae for the aberration coefficients can be conveniently expressed in terms of functions a_i b_i, c_i, d_i, e_i and f, and integrated terms $[a_i], \ldots, [f]$, defined as follows:

$$a_1 = (T_1 + T_2 + T_3)\, g_x^4\,; \quad a_2 = (T_1 - T_2 + T_3)\, g_y^4\,;$$

$$a_3 = (T_1 + T_2 + T_3)\, h_x^4\,; \quad a_4 = (T_1 - T_2 + T_3)\, h_y^4\,;$$

$$b_1 = 2(T_1 - T_3)\, g_x^2\, g_y^2 + 2T_4\, g_x\, g_y (g_x\, g_y' - g_x'\, g_y) + T_5 (g_x\, g_y' - g_x'\, g_y)^2\,;$$

$b_1 \to b_2$ if $g_y \to h_y$; $b_1 \to b_3$ if $g_x \to h_x$; $b_2 \to b_4$ if $g_x \to h_x$.

$$c_1 = 6(T_1 + T_2 + T_3)\, g_x^2\, h_x^2\,; \quad c_2 = 6(T_1 - T_2 + T_3)\, g_y^2\, h_y^2\,;$$

$$d_1 = 4(T_1 + T_2 + T_3)\, g_x^3\, h_x\,; \quad d_2 = 4(T_1 - T_2 + T_3)\, g_y^3\, h_y\,;$$

$$d_3 = 4(T_1 + T_2 + T_3)\, g_x\, h_x^3\,; \quad d_4 = 4(T_1 - T_2 + T_3)\, g_y\, h_y^3\,;$$

$$e_1 = 4(T_1 - T_3)\, g_x^2\, g_y\, h_y + 2T_4 \{g_x^2 (g_y\, h_y)' - g_x^{2\prime}\, g_y\, h_y\} + \\ + 2T_5 (g_x\, g_y' - g_x'\, g_y)(g_x\, h_y' - g_x'\, h_y)$$

$e_1 \to e_2$ if $g_x \leftrightarrow g_y$, $h_y \to h_x$, $D \to -D$, $Q \to -Q$;

$e_1 \to e_3$ if $g_x \to h_x$; $e_2 \to e_4$ if $g_y \to h_y$.

$$f = 8(T_1 - T_3)\, g_x\, h_x\, g_y\, h_y + 4T_4 \{g_x\, h_x (g_y\, h_y)' - (g_x\, h_x)'\, g_x\, h_x\} + \\ + 4T_5 \{2 (g_x\, h_x\, g_y'\, h_y' + g_x'\, h_x'\, g_y\, h_y) - (g_x\, h_x)'\, (g_y\, h_y)'\}$$

$$[a_1] = [(t_1 + t_2)\, g_x^4 + 2t_3\, g_x\, g_x'^3 + (4t_4 + t_5)\, g_x^2\, g_x'^2 + \\ + 2(t_6 + t_7 + t_8)\, g_x^3\, g_x']\,;$$

$[a_1] \to [a_2]$ if $D \to -D$, $Q \to -Q$, $g_x \to g_y$;

$[a_1] \to [a_3]$ if $g_x \to h_x$; $[a_3] \to [a_4]$ if $D \to -D$, $Q \to -Q$, $h_x \to h_y$

$$[b_1] = [2t_1\, g_x^2\, g_y^2 + 2t_3\, g_x'\, g_y' (g_x\, g_y)' + 8t_4\, g_x\, g_x'\, g_y\, g_y' + \\ + t_5 (g_x^2\, g_y'^2 + g_x'^2\, g_y^2) + 2t_6\, g_x\, g_y (g_x\, g_y)' + \\ + 2(t_7 - t_8)\, g_x\, g_y (g_x\, g_y' - g_x'\, g_y)]\,;$$

$[b_1] \to [b_2]$ if $g_y \to h_y$; $[b_1] \to [b_3]$ if $g_x \to h_x$; $[b_3] \to [b_4]$ if $g_y \to h_y$

$$[c_1] = [6(t_1 + t_2)\, g_x^2\, h_x^2 + 6t_3\, g_x'\, h_x' (g_x\, h_x)' + \\ + (4t_4 + t_5)(g_x^2\, h_x'^2 + 4g_x\, g_x'\, h_x\, h_x' + g_x'^2\, h_x^2) + \\ + 6(t_6 + t_7 + t_8)\, g_x\, h_x (g_x\, h_x)']\,;$$

$[c_1] \to [c_2]$ if $D \to -D$, $Q \to -Q$, $g_x \to g_y$, $h_x \to h_y$;

$$[d_1] = [4(t_1 + t_2)\, g_x^3 h_x + 2t_3\, g_x'^2 (3g_x h_x' + g_x' h_x) + $$
$$+ 8t_4\, g_x g_x' (g_x h_x)' + 2t_5\, g_x g_x' (g_x h_x)' + $$
$$+ 2(t_6 + t_7 + t_8)\, g_x^2 (g_x h_x' + 3g_x' h_x)];$$

$$[d_1] \to [d_2] \quad \text{if} \quad D \to -D, \quad Q \to -Q, \quad g_x \to g_y, \quad h_x \to h_y;$$

$$[d_1] \to [d_3] \quad \text{if} \quad g_x \leftrightarrow h_x; \quad [d_2] \to [d_4] \quad \text{if} \quad g_y \leftrightarrow h_y;$$

$$[e_1] = [4t_1\, g_x^2 g_y h_y + 2t_3 \{g_x^{2\prime} g_y' h_y' + g_x'^2 (g_y h_y)'\} + $$
$$+ 8t_4\, g_x g_x' (g_y h_y)' + 2t_5 (g_x^2 g_y' h_y' + g_x'^2 g_y h_y) + $$
$$+ 2t_6 \{g_x^2 (g_y h_y)' + g_x^{2\prime} g_y h_y\} + $$
$$+ 2(t_7 - t_8)\, \{g_x^2 (g_y h_y)' - g_x^{2\prime} g_y h_y\}];$$

$$[e_1] \to [e_2] \quad \text{if} \quad D \to -D, \quad Q \to -Q, \quad g_x \leftrightarrow g_y, \quad h_y \to h_x;$$

$$[e_1] \to [e_3] \quad \text{if} \quad g_x \to h_x; \quad [e_2] \to [e_4] \quad \text{if} \quad g_y \to h_y;$$

$$[f] = [8t_1\, g_x h_x g_y h_y + 4t_3 \{(g_x h_x)' g_y' h_y' + g_x' h_x' (g_y h_y)'\} + $$
$$+ 8t_4 (g_x h_x)' (g_y h_y)' + 4t_5 (g_x h_x g_y' h_y' + g_x' h_x' g_y h_y) + $$
$$+ 4(t_6 + t_7 - t_8)\, \{g_x h_x (g_y h_y)' + (g_x h_x)' g_y h_y\}]. \tag{3.38}$$

If the limits of integration lie in field-free space, all the t_i, with the exception of $t_3 = -\sqrt{\Phi}/16$, vanish; if the upper limit of integration is a stigmatic image point, only the following integrated terms are required:

$$[a_3] = 0; \quad [a_4] = 0; \quad [b_2] = \left[-\frac{\sqrt{\Phi}}{8} g_x g_x' h_y'^2\right]_o^i;$$

$$[b_3] = \left[-\frac{\sqrt{\Phi}}{8} g_y g_y' h_x'^2\right]_o^i; \quad [b_4] = 0;$$

$$[c_1] = \left[-\frac{3\sqrt{\Phi}}{8} g_x g_x' h_x'^2\right]_o^i; \quad [c_2] = \left[-\frac{3\sqrt{\Phi}}{8} g_y g_y' h_y'^2\right]_o^i;$$

$$[d_1] = \left[-\frac{3\sqrt{\Phi}}{8} g_x g_x'^2 h_x'\right]_o^i; \quad [d_2] = \left[-\frac{3\sqrt{\Phi}}{8} g_y g_y'^2 h_y'\right]_o^i;$$

$$[d_3] = \left[-\frac{\sqrt{\Phi}}{8} g_x h_x'^3\right]_o^i; \quad [d_4] = \left[-\frac{\sqrt{\Phi}}{8} g_y h_y'^3\right]_o^i;$$

$$[e_1] = \left[-\frac{\sqrt{\Phi}}{8} (g_x^{2\prime} g_y' h_y' + g_x'^2 g_y h_y')\right]_o^i;$$

$$[e_2] = \left[-\frac{\sqrt{\Phi}}{8} (g_x' g_y^{2\prime} h_x' + g_x g_y'^2 h_x')\right]_o^i;$$

$$[e_3] = \left[-\frac{\sqrt{\Phi}}{8} (g_y' h_x^{2\prime} h_y' + g_y h_x'^2 h_y')\right]_o^i;$$

$$[e_4] = \left[-\frac{\sqrt{\Phi}}{8} (g_x' h_x' h_y^{2\prime} + g_x h_x' h_y'^2)\right]_o^i;$$

$$[f] = \left[-\frac{\sqrt{\Phi}}{4} (g_x g_y)' h_x' h_y'\right]_o^i. \tag{3.39}$$

In terms of these quantities, then, we find that the aberration coefficients are as follows:

$$x_o^3:\ k_x(3000) = 4h_x\left(\int_a^c a_1\,dz + [a_1]_a^c\right) - g_x\left(\int_o^c d_1\,dz + [d_1]_o^c\right)$$

$$x_o^2 y_o:\ k_y(2100) = 2h_y\left(\int_a^c b_1\,dz + [b_1]_a^c\right) - g_y\left(\int_o^c e_1\,dz + [e_1]_o^c\right)$$

$$x_o y_o^2:\ k_x(1200) = 2h_x\left(\int_a^c b_1\,dz + [b_1]_a^c\right) - g_x\left(\int_o^c e_2\,dz + (e_2]_o^c\right)$$

$$y_o^3:\ k_y(0300) = 4h_y\left(\int_a^c a_2\,dz + [a_2]_a^c\right) - g_y\left(\int_o^c d_2\,dz + [d_2]_o^c\right)$$

$$x_a^3:\ k_x(0030) = h_x\left(\int_a^c d_3\,dz + [d_3]_a^c\right) - 4g_x\left(\int_o^c a_3\,dz + [a_3]_o^c\right)$$

$$x_a^2 y_a:\ k_y(0021) = h_y\left(\int_a^c e_3\,dz + [e_3]_a^c\right) - 2g_y\left(\int_o^c b_4\,dz + [b_4]_o^c\right)$$

$$x_a y_a^2:\ k_x(0012) = h_x\left(\int_a^c e_4\,dz + [e_4]_a^c\right) - 2g_x\left(\int_o^c b_4\,dz + [b_4]_o^c\right)$$

$$y_a^3:\ k_y(0003) = h_y\left(\int_a^c d_4\,dz + [d_4]_a^c\right) - 4g_y\left(\int_o^c a_4\,dz + [a_4]_o^c\right)$$

$$x_o^2 x_a:\ k_x(2010) = 3h_x\left(\int_a^c d_1\,dz + [d_1]_a^c\right) - 2g_x\left(\int_o^c c_1\,dz + [c_1]_o^c\right)$$

$$x_o^2 y_a:\ k_y(2001) = h_y\left(\int_a^c e_1\,dz + [e_1]_a^c\right) - 2g_y\left(\int_o^c b_2\,dz + [b_2]_o^c\right)$$

$$y_o^2 x_a:\ k_x(0210) = h_x\left(\int_a^c e_2\,dz + [e_2]_a^c\right) - 2g_x\left(\int_o^c b_3\,dz + [b_3]_o^c\right)$$

$$y_o^2 y_a:\ k_y(0201) = 3h_y\left(\int_a^c d_2\,dz + [d_2]_a^c\right) - 2g_y\left(\int_o^c c_2\,dz + [c_2]_o^c\right)$$

$$x_o x_a^2:\ k_x(1020) = 2h_x\left(\int_a^c c_1\,dz + [c_1]_a^c\right) - 3g_x\left(\int_o^c d_3\,dz + [d_3]_o^c\right)$$

$$y_o x_a^2:\ k_y(0120) = 2h_y\left(\int_a^c b_3\,dz + [b_3]_a^c\right) - g_y\left(\int_o^c e_3\,dz + [e_3]_o^c\right)$$

$$x_o y_a^2:\ k_x(1002) = 2h_x\left(\int_a^c b_2\,dz + [b_2]_a^c\right) - g_x\left(\int_o^c e_4\,dz + [e_4]_o^c\right)$$

$$y_o y_a^2:\ k_y(0102) = 2h_y\left(\int_a^c c_2\,dz + [c_2]_a^c\right) - 3g_y\left(\int_o^c d_4\,dz + [d_4]_o^c\right)$$

$$x_o y_o y_a:\ k_x(1101) = 2h_x\left(\int_a^c e_1\,dz + [e_1]_a^c\right) - g_x\left(\int_o^c f\,dz + [f]_o^c\right)$$

$$x_o y_o x_a\colon k_y(1110) = 2h_y\left(\int_a^c e_2\,\mathrm{d}z + [e_2]_a^c\right) - g_y\left(\int_o^c f\,\mathrm{d}z + [f]_o^c\right)$$

$$y_o x_a y_a\colon k_x(0111) = h_x\left(\int_a^c f\,\mathrm{d}z + [f]_a^c\right) - 2g_x\left(\int_o^c e_3\,\mathrm{d}z + [e_3]_o^c\right)$$

$$x_o x_a y_a\colon k_y(1011) = h_y\left(\int_a^c f\,\mathrm{d}z + [f]_a^c\right) - 2g_y\left(\int_o^c e_4\,\mathrm{d}z + [e_4]_o^c\right). \qquad (3.40)$$

Although these formulae are in the form appropriate to a system with an aperture, they can be very easily rewritten in the form required when the angular aperture at the object plane is bounded; from equations (3.11) and (3.12), it is clear that this entails replacing $g(z)$ by $s(z)$, $h(z)$ by $t(z)$, x_a and y_a by p_o and q_o respectively, and altering the lower limit of integration z_a to z_o wherever it occurs; k_x and k_y both reduce to unity.

The four aperture aberration coefficients can thus be written as follows (the object and current planes are assumed to lie in field free space, but the latter need not be a stigmatic image plane):

$$\begin{aligned} x_a^3\colon k_x(0030) = {}& 4h_x \int_a^c (T_1 + T_2 + T_3)\, g_x h_x^3\,\mathrm{d}z + \\ & + h_x[4(t_1 + t_2)\, g_x h_x^3 + 2t_3 h_x'^2 (g_x h_x' + 3g_x h_x') + \\ & + 8t_4 h_x h_x' (g_x h_x)' + 2t_5 h_x h_x' (g_x h_x)' + \\ & + 2(t_6 + t_7 + t_8)\, h_x^2 (2g_x h_x' + g_x' h_x)]_a^c - \\ & - 4g_x \int_o^c (T_1 + T_2 + T_3)\, h_x^4\,\mathrm{d}z + (g_x/2)\,[\sqrt{\Phi}\, h_x h_x'^3]_o^c . \end{aligned} \qquad (3.41\,\mathrm{a})$$

$$\begin{aligned} y_a^3\colon k_y(0003) = {}& 4h_y \int_a^c (T_1 - T_2 + T_3)\, g_y h_y^3\,\mathrm{d}z + \\ & + h_y[4(t_1 - t_2)\, g_y h_y^3 + 2t_3 h_y'^2 (g_y h_y' + 3g_y h_y') + \\ & + 8t_4 h_y h_y' (g_y h_y)' + 2t_5 h_y h_y' (g_y h_y)' + \\ & + 2(t_6 - t_7 - t_8)\, h_y^2 (3g_y h_y' + g_y' h_y)]_a^c - \\ & - 4g_y \int_o^c (T_1 - T_2 + T_3)\, h_y^4\,\mathrm{d}z + (g_y/2)\,[\sqrt{\Phi}\, h_y h_y'^3]_o^c . \end{aligned} \qquad (3.41\,\mathrm{b})$$

$$\begin{aligned} x_a y_a^2\colon k_x(0012) = {}& h_x \int_a^c [4(T_1 - T_3)\, g_x h_x h_y^2 - \\ & - 2T_4\{h_y^2 (g_x h_x)' - h_y^{2\prime} g_x h_x\} + \\ & + 2T_5 (g_x h_y' - g_x' h_y)(h_x h_y' - h_x' h_y)]\,\mathrm{d}z + \\ & + h_x[4t_1 g_x h_x h_y^2 + 2t_3\{g_x' h_x' h_y^{2\prime} + (g_x h_x)' h_y'^2\} + \\ & + 8t_4 (g_x h_x)' h_y h_y' + 2t_5 (g_x' h_x' h_y^2 + g_x h_x h_y'^2) + \\ & + 2t_6 (g_x h_x h_y^2)' + 2(t_7 - t_8)\{(g_x h_x)' h_y^2 - g_x h_x h_y^{2\prime}\}]_a^c - \\ & - 2g_x \int_o^c \{2(T_1 - T_3)\, h_x^2 h_y^2 + 2T_4 h_x h_y (h_x h_y' - h_x' h_y) + \\ & + T_5 (h_x h_y' - h_x' h_y)^2\}\,\mathrm{d}z + \\ & + (1/4)\, g_x [\sqrt{\Phi}\, h_x' h_y' (h_x h_y)']_o^c . \end{aligned} \qquad (3.41\,\mathrm{c})$$

$$x_a^2 y_a:\ k_y(0021) = h_y \int_a^c [4(T_1 - T_3)\, g_y\, h_x^2\, h_y +$$
$$+ 2T_4\{h_x^2 (g_y h_y)' - h_x^{2\prime} g_y h_y\} +$$
$$+ 2T_5(g_y' h_x - g_y h_x')(h_x h_y' - h_x' h_y)]\, dz +$$
$$+ h_y [4t_1 h_x^2 g_y h_y + 2t_3\{h_x^{2\prime} g_y' h_y' + h_x'^2 (g_y h_y)'\} +$$
$$+ 8t_4 h_x h_x' (g_y h_y)' + 2t_5 (h_x^2 g_y' h_y' + h_x'^2 g_y h_y) +$$
$$+ 2t_6 (h_x^2 g_y h_y)' + 2(t_7 - t_8)\{h_x^2 (g_y h_y)' - h_x^2 g_y h_y\}]_a^c -$$
$$- 2g_y \int_o^c \{2(T_1 - T_3) h_x^2 h_y^2 + 2T_4 h_x h_y (h_x h_y' - h_x' h_y) +$$
$$+ T_5 (h_x h_y' - h_x' h_y)^2\}\, dz +$$
$$+ (1/4)\, g_y [\sqrt{\Phi}\, h_x' h_y' (h_x h_y)']_o^c . \tag{3.41d}$$

When $V^{(4)}$ is given by equation (3.32) with $\Phi =$ constant, the aperture aberration coefficients involve the following quantities (both z_o and z_c lie in field-free space):

$$[a_3]_o^c = -(1/8)\,[\sqrt{\Phi}\, h_x h_x'^3]_o^c ;$$
$$[a_4]_o^c = -(1/8)\,[\sqrt{\Phi}\, h_y h_y'^3]_o^c ;$$
$$[b_4]_o^c = -(1/8)\,[\sqrt{\Phi}\, h_x' h_y' (h_x h_y)']_o^c$$
$$[d_3]_a^c = \left[-\frac{1}{8}\sqrt{\Phi}\, h_x'^2 (g_x h_x' + 3g_x' h_x) + \frac{5D - 12\eta Q\sqrt{\Phi}}{96\sqrt{\Phi}}\, g_x' h_x^3\right]_a^c$$
$$[d_4]_a^c = \left[-\frac{1}{8}\sqrt{\Phi}\, h_y'^2 (g_y h_y' + 3g_y' h_y) - \frac{5D - 12\eta Q\sqrt{\Phi}}{96\sqrt{\Phi}}\, g_y' h_y^3\right]_a^c$$
$$[e_3]_a^c = \left[-\frac{1}{8}\sqrt{\Phi}\, \{h_x^{2\prime} g_y' h_y' + h_x'^2 (g_y h_y)'\} + \frac{3D - 4\eta Q\sqrt{\Phi}}{32\sqrt{\Phi}}\, g_y' h_x^2 h_y\right]_a^c$$
$$[e_4]_a^c = \left[-\frac{1}{8}\sqrt{\Phi}\, \{h_y^{2\prime} g_x' h_x' + h_y'^2 (g_x h_x)'\} - \frac{3D - 4\eta Q\sqrt{\Phi}}{32\sqrt{\Phi}}\, g_x' h_x h_y^2\right]_a^c$$
$$a_3 = -\frac{D^2 - 4\eta DQ\sqrt{\Phi} + 6\eta^2 Q^2 \Phi}{48\,\Phi^{3/2}}\, h_x^4 - \frac{D' - 4\eta Q'\sqrt{\Phi}}{96\sqrt{\Phi}}\, h_x^3 h_x' ;$$
$$a_4 = -\frac{D^2 - 4\eta DQ\sqrt{\Phi} + 6\eta^2 Q^2 \Phi}{48\,\Phi^{3/2}}\, h_y^4 + \frac{D' - 4\eta Q'\sqrt{\Phi}}{96\sqrt{\Phi}}\, h_y^3 h_y' ;$$
$$b_4 = \frac{(D - 2\eta Q\sqrt{\Phi})^2}{16\,\Phi^{3/2}}\, h_x^2 h_y^2 + \frac{3}{32}\,\frac{D' - 4\eta Q'\sqrt{\Phi}}{\sqrt{\Phi}}\, h_x h_y (h_x' h_y - h_x h_y') ;$$
$$d_3 = -\frac{D^2 - 4\eta DQ\sqrt{\Phi} + 6\eta^2 Q^2 \Phi}{12\,\Phi^{3/2}}\, g_x h_x^3 - \frac{D' - 4\eta Q'\sqrt{\Phi}}{96\sqrt{\Phi}}\, (g_x h_x^3)' ;$$
$$d_4 = -\frac{D^2 - 4\eta DQ\sqrt{\Phi} + 6\eta^2 Q^2 \Phi}{12\,\Phi^{3/2}}\, g_y h_y^3 + \frac{D' - 4\eta Q'\sqrt{\Phi}}{96\sqrt{\Phi}}\, (g_y h_y^3)'$$
$$e_3 = \frac{(D - 2\eta Q\sqrt{\Phi})^2}{8\,\Phi^{3/2}}\, h_x^2 g_y h_y +$$
$$+ \frac{3}{32}\,\frac{D' - 4\eta Q'\sqrt{\Phi}}{\sqrt{\Phi}}\, \{h_x^{2\prime} g_y h_y - h_x^2 (g_y h_y)'\}$$
$$e_4 = \frac{(D - 2\eta Q\sqrt{\Phi})^2}{8\,\Phi^{3/2}}\, g_x h_x h_y^2 +$$
$$+ \frac{3}{32}\,\frac{D' - 4\eta Q'\sqrt{\Phi}}{\sqrt{\Phi}}\, \{(g_x h_x)' h_y^2 - g_x h_x h_y^{2\prime}\} . \tag{3.42}$$

The aperture aberration coefficients corresponding to $V^{(4)}$ in the form (3.34) contain no derivatives of D and Q, thus:

$$[a_3] = \left[-\frac{D' - 2\eta\, Q' \sqrt{\Phi}}{96\sqrt{\Phi}}\, h_x^4 - \frac{1}{12}\sqrt{\Phi}\, h_x\, h_x'^3 + \frac{D - 2\eta\, Q\sqrt{\Phi}}{24\sqrt{\Phi}}\, h_x^3\, h_x'\right];$$

$$[a_4] = \left[\frac{D' - 2\eta\, Q' \sqrt{\Phi}}{96\sqrt{\Phi}}\, h_y^4 - \frac{1}{12}\sqrt{\Phi}\, h_y\, h_y'^3 - \frac{D - 2\eta\, Q\sqrt{\Phi}}{24\sqrt{\Phi}}\, h_y^3\, h_y'\right];$$

$$[b_4] = \left[-\frac{1}{12}\sqrt{\Phi}\, h_x'\, h_y' (h_x\, h_y)' + \frac{1}{4}\eta\, Q\, h_x\, h_y (h_x\, h_y' - h_x'\, h_y)\right];$$

$$[d_3] = \left[-\frac{D' - 2\eta\, Q'\sqrt{\Phi}}{24\sqrt{\Phi}}\, g_x\, h_x^3 - \frac{1}{12}\sqrt{\Phi}\, h_x'^2 (g_x\, h_x' + 3 g_x'\, h_x) + \right.$$
$$\left. + \frac{D - 2\eta\, Q\sqrt{\Phi}}{24\sqrt{\Phi}}\, h_x^2 (3 g_x\, h_x' + g_x'\, h_x)\right];$$

$$[d_4] = \left[\frac{D' - 2\eta\, Q'\sqrt{\Phi}}{24\sqrt{\Phi}}\, g_y\, h_y^3 - \frac{1}{12}\sqrt{\Phi}\, h_y'^2 (g_y\, h_y' + 3 g_y'\, h_y) - \right.$$
$$\left. - \frac{D - 2\eta\, Q\sqrt{\Phi}}{24\sqrt{\Phi}}\, h_y^2 (3 g_y\, h_y' + g_y'\, h_y)\right];$$

$$[e_3] = \left[-\frac{1}{12}\sqrt{\Phi}\, \{h_x^{2\prime}\, g_y'\, h_y' + h_x'^2 (g_y\, h_y)'\} + \right.$$
$$\left. + \frac{1}{4}\eta\, Q\, \{h_x^2 (g_y\, h_y)' - h_x^{2\prime}\, g_y\, h_y\}\right];$$

$$[e_4] = \left[-\frac{1}{12}\sqrt{\Phi}\, \{h_y^{2\prime}\, g_x'\, h_x' + h_y'^2 (g_x\, h_x)'\} - \right.$$
$$\left. - \frac{1}{4}\eta\, Q \{h_y^2 (g_x\, h_x)' - h_y^{2\prime}\, g_x\, h_x\}\right].$$

$$a_3 = -\frac{7D^2 - 24\eta\, DQ\sqrt{\Phi} + 32\eta^2 Q^2 \Phi}{384\,\Phi^{3/2}}\, h_x^4 - \frac{1}{24}\sqrt{\Phi}\, h_x'^4$$

$$a_4 = -\frac{7D^2 - 24\eta\, DQ\sqrt{\Phi} + 32\eta^2 Q^2 \Phi}{384\,\Phi^{3/2}}\, h_y^4 - \frac{1}{24}\sqrt{\Phi}\, h_y'^4$$

$$b_4 = \frac{D - 4\eta\, Q\sqrt{\Phi}}{12\sqrt{\Phi}}\, (h_x^2 h_y'^2 - h_x'^2\, h_y^2) - \frac{1}{12}\sqrt{\Phi}\, h_x'^2\, h_y'^2 +$$
$$+ \frac{D^2 + 8\eta\, DQ\sqrt{\Phi} - 32\eta^2 Q^2 \Phi}{64\,\Phi^{3/2}}\, h_x^2\, h_y^2$$

$$d_3 = -\frac{7D^2 - 24\eta\, DQ\sqrt{\Phi} + 32\eta^2 Q^2 \Phi}{96\,\Phi^{3/2}}\, g_x\, h_x^3 - \frac{1}{6}\sqrt{\Phi}\, g_x'\, h_x'^3$$

$$d_4 = -\frac{7D^2 - 24\eta\, DQ\sqrt{\Phi} + 32\eta^2 Q^2 \Phi}{96\,\Phi^{3/2}}\, g_y\, h_y^3 - \frac{1}{6}\sqrt{\Phi}\, g_y'\, h_y'^3$$

$$e_3 = \frac{D^2 + 8\eta\, DQ\sqrt{\Phi} - 32\eta^2 Q^2 \Phi}{32\,\Phi^{3/2}}\, h_x^2 g_y\, h_y - \frac{1}{6}\sqrt{\Phi}\, h_x'^2\, g_y'\, h_y' +$$
$$+ \frac{D - 4\eta\, Q\sqrt{\Phi}}{6\sqrt{\Phi}}\, (h_x^2\, g_y'\, h_y' - h_x'^2\, g_y\, h_y);$$

$$e_4 = \frac{D^2 + 8\eta\, DQ\sqrt{\Phi} - 32\eta^2 Q^2 \Phi}{32\,\Phi^{3/2}}\, g_x\, h_x\, h_y^2 - \frac{1}{6}\sqrt{\Phi}\, g_x'\, h_x'\, h_y'^2 +$$
$$+ \frac{D - 4\eta\, Q\sqrt{\Phi}}{6\sqrt{\Phi}}\, (g_x\, h_x\, h_y'^2 - g_x'\, h_x'\, h_y^2)\,. \qquad (3.43)$$

In a stigmatic system, the integrated terms $[a_3]$, $[a_4]$ and $[b_4]$ vanish; of the three integrands a_3, a_4 and b_4, a_3 and a_4 are never positive, so that the signs of the aberration coefficients (0030) and (0003) of a stigmatic quadrupole system are the same as those of g_{xi} and g_{yi} respectively [143, 95, 140b]. By setting $\varrho_1 = 16$ in equation (3.34), it is easy to show that b_4 is never positive for magnetic quadrupoles [143, 140b].

3.3. Integral equations

The equations of motion, including terms arising from $m^{(4)}$ [equations (3.1)], can be transformed into integral equations by constructing the appropriate Green's functions. This technique has been explored in detail by *Meads* [130, 132] for the case of magnetic quadrupoles ($D = 0$, $\Phi =$ constant). In this case, the equations of motion simplify to

$$x'' + \mathfrak{Q}\, x = {}^1\!/_{12}\, \mathfrak{Q}''\, x^3 + {}^1\!/_4\, \mathfrak{Q}''\, x\, y^2 + \\ + \mathfrak{Q}'\, x\, y\, y' - {}^3\!/_2\, \mathfrak{Q}\, x\, x'^2 - {}^1\!/_2\, \mathfrak{Q}\, x\, y'^2 + \mathfrak{Q}\, y\, x'\, y' \tag{3.44a}$$

$$y'' - \mathfrak{Q}\, y = -{}^1\!/_{12}\, \mathfrak{Q}''\, y^3 - {}^1\!/_4\, \mathfrak{Q}''\, x^2\, y - \\ - \mathfrak{Q}'\, x\, y\, x' + {}^3\!/_2\, \mathfrak{Q}\, y\, y'^2 + {}^1\!/_2\, \mathfrak{Q}\, y\, x'^2 - \mathfrak{Q}\, x\, x'\, y' \tag{3.44b}$$

in which $\mathfrak{Q}$ denotes $\eta\, Q/\sqrt{\Phi}$. Although *Meads'* work was performed with strong-focusing lenses in mind, his analysis is deliberately set out in such

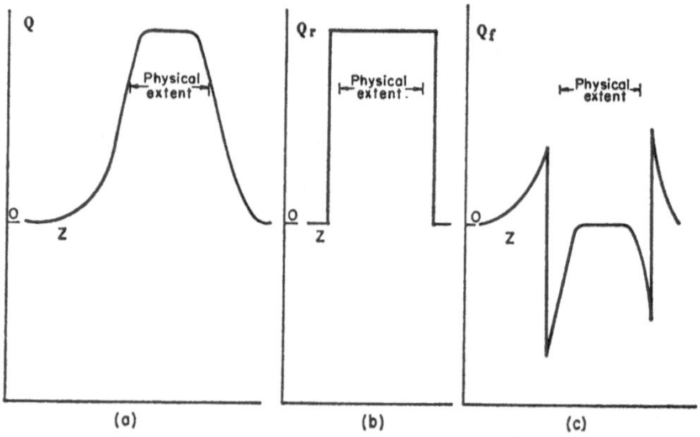

Fig. 5a—c. *Meads'* division of the potential distribution in a long quadrupole (a) into rectangular distribution (b) together with fringe potential (c). (By courtesy of Dr. *Meads* and the University of California Lawrence Radiation Laboratory, Berkeley)

a way that the quadrupoles of lower energy systems (microscopes and microanalysers, for example) are included. In strong-focusing lenses, the potential distribution $\mathfrak{Q}$ can be regarded as a rectangular distribution (suffix r) together with a fringing field (suffix f); thus

$$\mathfrak{Q} = \mathfrak{Q}_r + \mathfrak{Q}_f$$

(see Fig. 5) and the width of $\mathfrak{Q}_r$ is chosen such that $\int \mathfrak{Q}_f\, \mathrm{d}z$ vanishes when $\mathfrak{Q}_f$ and $\mathfrak{Q}_f'$ vanish at the end-points of the integral; $\mathfrak{Q}_f$ is assumed to vanish well inside the quadrupole. The solution of equation (3.44a) is then written in the form

$$x = x_r + x_f + x^{(3)} \tag{3.45}$$

in which

$$x_r'' + \mathfrak{Q}_r x_r = 0 \tag{3.46a}$$

$$x_f'' + \mathfrak{Q}_r x_f = -\mathfrak{Q}_f (x_r + x_f) + \Delta \mathfrak{Q} (x_f + x_c) \tag{3.46b}$$

$$\begin{aligned} x^{(3)''} + \mathfrak{Q}\, x^{(3)} = {}^1/_{12}\, \mathfrak{Q}''\, x^3 + {}^1/_4\, \mathfrak{Q}''\, x\, y^2 + \mathfrak{Q}'\, x\, y\, y' - \\ - {}^3/_2\, \mathfrak{Q}\, x\, x'^2 - {}^1/_2\, \mathfrak{Q}\, x\, y'^2 + \mathfrak{Q}\, y\, x'\, y'\,. \end{aligned} \tag{3.46c}$$

We shall speak of this as the long-quadrupole case*. The term in Δ allows the chromatic aberration to be calculated simultaneously; Δ denotes $\sqrt{(\Phi_1/\Phi_0)} - 1$ if the accelerating voltage Φ_0 is perturbed to Φ_1.

For long quadrupoles, we define two Green's functions, satisfying the differential equations

$$\begin{aligned} \frac{\mathrm{d}^2\, \mathfrak{g}(z, \zeta)}{\mathrm{d} z^2} + \mathfrak{Q}_r(z)\, \mathfrak{g}(z, \zeta) = -\delta(z - \zeta) \\ \frac{\mathrm{d}^2\, \mathfrak{g}_f(z, \zeta)}{\mathrm{d} z^2} + \mathfrak{Q}\, \mathfrak{g}_f(z, \zeta) = -\delta(z - \zeta)\,. \end{aligned} \tag{3.47a}$$

The linearly independent solutions of (3.46a) that we use are those appropriate to a system without an aperture, $s_x^{(r)}$, $t_x^{(r)}$, $s_y^{(r)}$ and $t_y^{(r)}$ (with boundary conditions $s_{xo}^{(r)} = 1$, $s_{xo}^{(r)\prime} = 0$, $t_{xo}^{(r)} = 0$, $t_{xo}^{(r)\prime} = 1/\sqrt{\Phi_o}$) and since we are considering only quadrupole systems, we write

$$x_r = s_x^{(r)}\, x_o + T_x^{(r)}\, x_o'\,, \quad y_r = s_y^{(r)}\, y_o + T_y^{(r)}\, y_o'$$

with $T^{(r)} = \sqrt{\Phi_o}\, t^{(r)}$. Thus, for

$$\zeta \leqq z\,, \quad \mathfrak{g}(z, \zeta) = T_x^{(r)}(z)\, s_x^{(r)}(\zeta) - s_x^{(r)}(z)\, T_x^{(r)}(\zeta)$$

and for

$$\zeta \geqq z\,, \quad \mathfrak{g}(z, \zeta) = 0\,. \tag{3.47b}$$

The corresponding solutions of the equation $x'' + \mathfrak{Q}\, x = 0$ are written s_x, T_x and the Green's function is thus

$$\mathfrak{g}_f(z, \zeta) = T_x(z)\, s_x(\zeta) - s_x(z)\, T_x(\zeta) \quad \text{for} \quad \zeta \leqq z\,. \tag{3.47c}$$

In terms of these functions, the differential equations (3.46b and c) become the integral equations

$$x_f(z) = \int_o^z \mathfrak{g}(z, \zeta)\, \{x_r(\zeta) + x_f(\zeta)\}\, \{-\mathfrak{Q}_f(\zeta) + \Delta\, \mathfrak{Q}(\zeta)\}\, \mathrm{d}\zeta \tag{3.48a}$$

$$\begin{aligned} x^{(3)}(z) = \int_o^z \mathfrak{g}_f(z, \zeta)\, ({}^1/_{12}\, \mathfrak{Q}''\, x^3 + {}^1/_4\, \mathfrak{Q}''\, x\, y^2 + \mathfrak{Q}'\, x\, y\, y' - \\ - {}^3/_2\, \mathfrak{Q}\, x\, x'^2 - {}^1/_2\, \mathfrak{Q}\, x\, y'^2 + \mathfrak{Q}\, y\, x'\, y')\, \mathrm{d}\zeta \end{aligned} \tag{3.48b}$$

(in the equation for $x^{(3)}$, $\mathfrak{Q}$ is a function of ζ and $\mathfrak{Q}' = \mathrm{d}\mathfrak{Q}/\mathrm{d}\zeta, \ldots$).

* A "long" quadrupole is of course, one in which the length is much greater than the distance between each pole and the axis.

The chromatic aberrations, and the primary aberrations which arise from the fringe field, are given by equations (3.48). The function x_f is given by

$$x_f = I_a + I_b + I_c + I_d + I_e + I_f \tag{3.49}$$

where

$$I_a = -\int_0^z \mathfrak{g}(z,\zeta)\,\mathfrak{Q}_f(\zeta)\,x_r(\zeta)\,\mathrm{d}\zeta = -\int_0^z \{T_x^{(r)}(z)\,s_x^{(r)}(\zeta) - - s_x^{(r)}(z)\,T_x^{(r)}(\zeta)\}\,\mathfrak{Q}_f\,x_r\,\mathrm{d}\zeta$$

$$I_b = \Delta\int_0^z \mathfrak{g}(z,\zeta)\,\mathfrak{Q}_r(\zeta)\,x_r(\zeta)\,\mathrm{d}\zeta$$

$$I_c = -\int_0^z \mathfrak{g}(z,\zeta)\,\mathfrak{Q}_f(\zeta)\,x_f(\zeta)\,\mathrm{d}\zeta$$

$$I_d = \Delta\int_0^z \mathfrak{g}(z,\zeta)\,\mathfrak{Q}_r(\zeta)\,x_f(\zeta)\,\mathrm{d}\zeta$$

$$I_e = \Delta\int_0^z \mathfrak{g}(z,\zeta)\,\mathfrak{Q}_f(\zeta)\,x_r(\zeta)\,\mathrm{d}\zeta$$

$$I_f = \Delta\int_0^z \mathfrak{g}(z,\zeta)\,\mathfrak{Q}_f(\zeta)\,x_f(\zeta)\,\mathrm{d}\zeta\,. \tag{3.50}$$

If h is some length, characteristic of the breadth of the fringe field, we can define two parameters: ε, which is a measure of the slope of the rays and λ, defined as $h/\sqrt{\mathfrak{Q}_r}$; to obtain primary aberrations, we neglect terms higher than third order in Δ, λ and ε. It can be shown that

$$I_a = \sum_k \left(-c_k\,\lambda_k^2\,\mathfrak{Q}_{rk}\right)\left[\frac{\partial}{\partial\zeta}\{\mathfrak{g}(z,\zeta)\,x_r(\zeta)\}\right]_{\zeta=z_k}$$

$$I_b = \Delta\int_0^z \mathfrak{g}(z,\zeta)\,\mathfrak{Q}_r(\zeta)\,x_r(\zeta)\,\mathrm{d}\zeta$$

$$I_c = O(\Delta\,\lambda^2\,\varepsilon)$$

$$I_d = \Delta^2\int_0^z\left\{\mathfrak{Q}_r(\zeta)\,\mathfrak{g}(z,\zeta)\int^{\zeta}\mathfrak{Q}_r(\xi)\,\mathfrak{g}(\zeta,\xi)\,\mathrm{d}\xi\right\}\mathrm{d}\zeta$$

$$I_e = O(\Delta\,\lambda^2\,\varepsilon)$$

$$I_f = O(\Delta^2\,\lambda^2\,\varepsilon) \tag{3.51}$$

so that

$$x_f = I_a + I_b + I_d\,.$$

The summations are to be taken over each fringe zone (k) and the coefficients $c_k\,\lambda_k^2$ are defined by

$$c_k\,\lambda_k^2 = (1/\mathfrak{Q}_r)\int\xi\,\mathfrak{Q}_f(\xi)\,\mathrm{d}\xi \cong h^2$$

where the integration is taken over the fringe region.

The primary aberrations are derived from equation (3.48b). If the object and current planes lie outside the influence of the quadrupole fields, elimination of $\mathfrak{Q}'$ and $\mathfrak{Q}''$ by partial integration yields

$$x^{(3)} = -\int_o^z \mathfrak{Q}^2\,\mathfrak{g}\left(\frac{1}{3}x^3 + x\,y^2\right)\mathrm{d}\zeta + \frac{1}{2}\int_o^z \mathfrak{Q}\,\mathfrak{g}'\,x'(x^2+y^2)\,\mathrm{d}\zeta - \\ - \int_o^z \mathfrak{Q}\,\mathfrak{g}(x\,x'^2 + x\,y'^2 - y\,x'\,y')\,\mathrm{d}\zeta \qquad (3.52)$$

in which $\mathfrak{g}$ is given by equation (3.47c).

The aberration coefficients can be expressed in a very succinct fashion by relabelling the functions s and T and their derivatives, and the boundary conditions. Following *Meads*, we set

$$\begin{array}{llll} \chi_1 = s_x^{(r)} & \chi_2 = s_y^{(r)} & \chi_3 = T_x^{(r)} & \chi_4 = T_y^{(r)} \\ \chi_5 = s_x^{(r)\prime} & \chi_6 = s_y^{(r)\prime} & \chi_7 = T_x^{(r)\prime} & \chi_8 = T_y^{(r)\prime} \end{array} \qquad (3.53\mathrm{a})$$

and

$$\begin{array}{llll} \psi_1 = x_o & \psi_2 = y_o & \psi_3 = x_o' & \psi_4 = y_o' \\ \psi_5 = \Delta\,. \end{array} \qquad (3.53\mathrm{b})$$

It is convenient to write $\psi_0 = \psi_6 = 1$.

The contribution to $x^{(3)}$ and $y^{(3)}$ which arises from the rectangular field distribution consists of terms of the form

$$\psi_i\,\psi_j\,\psi_k \int_o^z \mathfrak{Q}_r^2\,\chi_m(\zeta)\,\chi_n(\zeta)\,\chi_p(\zeta)\,\chi_q(\zeta)\,\mathrm{d}\zeta$$

and

$$\psi_i\,\psi_j\,\psi_k \int_o^z \mathfrak{Q}_r\,\chi_m(\zeta)\,\chi_n(\zeta)\,\chi_{p+4}(\zeta)\,\chi_{q+4}(\zeta)\,\mathrm{d}\zeta$$

in which m, n, p, q are all $\leqq 4$. The integrals in these expressions are denoted by $(m\,n\,p\,q)$:

$$(m\,n\,p\,q) = \int_o^z \mathfrak{Q}_r^2\,\chi_m\,\chi_n\,\chi_p\,\chi_q\,\mathrm{d}\zeta \quad (m, n, p, q \quad \text{all} \quad \leqq 4)$$

or

$$(m\,n\,p\,q) = \int_o^z \mathfrak{Q}_r\,\chi_m\,\chi_n\,\chi_p\,\chi_q\,\mathrm{d}\zeta \quad (m, n, p, q \quad \text{not all} \quad \leqq 4)\,.$$

The function x_f contains integrals of the form

$$(m\,n) = \int_o^z \mathfrak{Q}_r(\zeta)\,\chi_m(\zeta)\,\chi_n(\zeta)\,\mathrm{d}\zeta$$

and

$$(m\,n;\,p\,q) = \int_o^z \left\{\mathfrak{Q}_r(\zeta)\,\chi_m(\zeta)\,\chi_n(\zeta)\int_o^\zeta \mathfrak{Q}_r\,\chi_p(\xi)\,\chi_q(\xi)\,\mathrm{d}\xi\right\}\mathrm{d}\zeta$$

and sums over the fringe fields of the form

$$S_{mn} = \sum_k c_k\,\mathfrak{Q}_{rk}\,\chi_m(z_k)\,\chi_n(z_k)\,.$$

The combined effect of x_f and $x^{(3)}$ is thus of the form

$$x_f(z) + x^{(3)}(z) = \sum_{1 \leq k \leq j \leq i \leq 6} (i\,j\,k)\, \psi_i\, \psi_j\, \psi_k \,.$$

The coefficients $(i\,j\,k)$ are given by the following expressions, which correspond to those calculated by means of the object characteristic, when the form of $\int m^{(4)}\,\mathrm{d}z$ free of derivatives of Ω is selected, and $D = 0$, $\Phi =$ constant. (We recall that x_o' and y_o' in the formulae below may be replaced by $p_o/\sqrt{\Phi}$ and $q_o/\sqrt{\Phi}$.) The coefficients $(\alpha\,\beta\,\gamma\,\delta)$ which are listed are those used when we write the aberrations in the form $(\alpha\,\beta\,\gamma\,\delta)\, x_o\, y_o\, x_o'\, y_o'$; the index (r) is dropped.

$$(0003)\; y_o'^3\!:\; (444) = s_y\left\{\frac{1}{3}(4444) - \frac{1}{2}(8844) + \frac{1}{3}\langle 4444\rangle\right\} + $$
$$+ T_y\left\{-\frac{1}{3}(4442) - \frac{1}{2}(8644) + (8842) - \frac{1}{3}\langle 4442\rangle\right\}$$

$$(0030)\; x_o'^3\!:\; (333) = s_x\left\{\frac{1}{3}(3333) + \frac{1}{2}(7733) + \frac{1}{3}\langle 3333\rangle\right\} + $$
$$+ T_x\left\{-\frac{1}{3}(3331) + \frac{1}{2}(7533) - (7731) - \frac{1}{3}\langle 3331\rangle\right\}$$

$$(0300)\; y_o^3\!:\; (222) = s_y\left\{\frac{1}{3}(4222) + \frac{1}{2}(8622) - (6642) + \frac{1}{3}\langle 4222\rangle\right\} + $$
$$+ T_y\left\{-\frac{1}{3}(2222) + \frac{1}{2}(6622) - \frac{1}{3}\langle 2222\rangle\right\}$$

$$(3000)\; x_o^3\!:\; (111) = s_x\left\{\frac{1}{3}(3111) - \frac{1}{2}(7511) + (5531) + \frac{1}{3}\langle 3111\rangle\right\} + $$
$$+ T_x\left\{-\frac{1}{3}(1111) - \frac{1}{2}(5511) - \frac{1}{3}\langle 1111\rangle\right\}$$

$$(0201)\; y_o^2\, y_o'\!:\; (422) = s_y\left\{(4422) - (8642) + \frac{1}{2}(8822) - (6644) + \langle 4422\rangle\right\} + $$
$$+ T_y\left\{-(4222) + \frac{3}{2}(8622) - \langle 4222\rangle\right\}$$

$$(2010)\; x_o^2\, x_o'\!:\; (311) = s_x\left\{(3311) + (7531) - \frac{1}{2}(7711) + (5533) + \langle 3311\rangle\right\} + $$
$$+ T_x\left\{-(3111) - \frac{3}{2}(7511) - \langle 3111\rangle\right\}. \tag{3.54}$$

$$(0102)\; y_o\, y_o'^2\!:\; (442) = s_y\left\{(4442) - \frac{3}{2}(8644) + \langle 4442\rangle\right\} + $$
$$+ T_y\left\{-(4422) - \frac{1}{2}(6644) + (8642) + (8822) - \langle 4422\rangle\right\}$$

$$(1020)\; x_o\, x_o'^2\!:\; (331) = s_x\left\{(3331) + \frac{3}{2}(7533) + \langle 3331\rangle\right\} + $$
$$+ T_x\left\{-(3311) + \frac{1}{2}(5533) - (7531) - (7711) - \langle 3311\rangle\right\}$$

$$(0021)\; x_o'^2\, y_o'\!:\; (433) = s_y\left\{(4433) + \frac{1}{2}(8843) - (7744) + (8743) - \langle 4433\rangle\right\} + $$
$$+ T_y\left\{-(4332)\,\frac{1}{2} - (8633) + (7742) - (8732) + \langle 4332\rangle\right\}$$

$$(0012)\; x_o' y_o'^2\colon (443) = s_x\left\{(4433) - \frac{1}{2}(7744) + (8833) - (8743) - \langle 4433\rangle\right\} +$$
$$+ T_x\left\{-(4431) + \frac{1}{2}(7544) - (8831) + (8741) + \langle 4431\rangle\right\}$$

$$(2100)\; x_o^2 y_o\colon (211) = s_y\left\{(4211) + \frac{1}{2}(8611) - (5542) + (6541) - \langle 4211\rangle\right\} +$$
$$+ T_y\left\{-(2211) - \frac{1}{2}(6611) + (5522) - (6521) + \langle 2211\rangle\right\}$$

$$(1200)\; x_o y_o^2\colon (221) = s_x\left\{(3221) - \frac{1}{2}(7522) + (6631) - (6532) - \langle 3221\rangle\right\} +$$
$$+ T_x\left\{-(2211) + \frac{1}{2}(5522) - (6611) + (6521) + \langle 2211\rangle\right\}$$

$$(2001)\; x_o^2 y_o'\colon (411) = s_y\left\{(4411) + \frac{1}{2}(8811) - (5544) + (8541) - \langle 4411\rangle\right\} +$$
$$+ T_y\left\{-(4211) - \frac{1}{2}(8611) + (5542) - (8521) + \langle 4211\rangle\right\}$$

$$(0210)\; y_o^2 x_o'\colon (322) = s_x\left\{(3322) - \frac{1}{2}(7722) + (6633) - (7632) - \langle 3322\rangle\right\} +$$
$$+ T_x\left\{-(3221) + \frac{1}{2}(7522) - (6631) + (7621) + \langle 3221\rangle\right\}$$

$$(0120)\; y_o x_o'^2\colon (332) = s_y\left\{(4332) + \frac{1}{2}(8633) - (7742) + (7643) - \langle 4332\rangle\right\} +$$
$$+ T_y\left\{-(3322) - \frac{1}{2}(6633) + (7722) - (7632) + \langle 3322\rangle\right\}$$

$$(1002)\; x_o y_o'^2\colon (441) = s_x\left\{(4431) - \frac{1}{2}(7544) + (8831) - (8543) - \langle 4431\rangle\right\} +$$
$$+ T_x\left\{-(4411) + \frac{1}{2}(5544) - (8811) + (8541) + \langle 4411\rangle\right\}$$

$$(1110)\; x_o y_o x_o'\colon (321) = s_y\{2(4321) + (8631) - 2(7542) + (6543) +$$
$$+ (7641) - 2\langle 4321\rangle\}$$
$$+ T_y\{-2(3221) - (6631) + 2(7522) - (6532) - (7621) + 2\langle 3221\rangle\}$$

$$(1101)\; x_o y_o y_o'\colon (421) = s_x\{2(4321) - (7542) + 2(8631) - (6543) -$$
$$- (8532) - 2\langle 4321\rangle\} +$$
$$+ T_x\{-2(4211) + (5542) - 2(8611) + (6541) + (8521) + 2\langle 4211\rangle\}$$

$$(1011)\; x_o x_o' y_o'\colon (431) = s_y\{-2(4431) + (8831) - 2(7544) + (8543) +$$
$$+ (8741) - 2\langle 4431\rangle\} +$$
$$+ T_y\{2(4321) - (8631) + 2(7542) - (8532) - (8721) + 2\langle 4321\rangle\}$$

$$(0111)\; y_o x_o' y_o'\colon (432) = s_x\{2(4332) - (7742) + 2(8633) - (7643) -$$
$$- (8732) - 2\langle 4332\rangle\} +$$
$$+ T_x\{-2(4321) + (7542) - 2(8631) + (7641) + (8721) + 2\langle 4321\rangle\}$$

$\langle m\, n\, p\, q\rangle$ denotes $\int \mathfrak{Q}_1(\zeta)\, \chi_m(\zeta)\, \chi_n(\zeta)\, \chi_p(\zeta)\, \chi_q(\zeta)\, d\zeta$ and $\mathfrak{Q}_1 = 12\eta\, Q_1/\sqrt{\Phi}$.

Chromatic aberrations:

$$\Delta x_o:\ (51) = T_x(11) - s_x(31)$$
$$\Delta y_o:\ (52) = -T_y(22) + s_y(42)$$
$$\Delta x_o':\ (53) = T_x(31) - s_x(33)$$
$$\Delta y_o':\ (54) = -T_y(42) + s_y(44)$$
$$\Delta^2 x_o:\ (551) = T_x\{(31;11) - (11;31)\} + s_x\{(31;31) - (33;11)\}$$
$$\Delta^2 y_o:\ (552) = T_y\{(42;22) - (22;42)\} + s_y\{(42;42) - (44;22)\}$$
$$\Delta^2 x_o':\ (553) = T_x\{(31;31) - (11;33)\} + s_x\{(31;33) - (33;31)\}$$
$$\Delta^2 y_o':\ (554) = T_y\{(42;42) - (22;44)\} + s_y\{(42;44) - (44;42)\}\,.$$

Fringe field sums:

$$x_o:\ (661) = 2\,T_x\,S_{51} - s_x(S_{71} + S_{53})$$
$$y_o:\ (662) = -2\,T_y\,S_{62} + s_y(S_{82} + S_{64})$$
$$x_o':\ (663) = T_x(S_{71} + S_{53}) - 2s_x\,S_{73}$$
$$y_o':\ (664) = -T_y(S_{82} + S_{64}) + 2s_y\,S_{84}\,.$$

Twenty coefficients are obtained for the x-component of the primary geometrical aberration and twenty will likewise be obtained for the y-component. A further thirty-two characterize the chromatic aberration coefficients. Twenty-four of the geometrical aberration coefficients are otiose however, and the corresponding relations between the coefficients can be derived by exploiting the invariance of the Poincaré integral. This superfluity is not a drawback of Meads' method, for these inter-relations provide an efficient method of checking the computed values*. The formulae, the derivation of which is given in full in [130], are as follows:

$$M_x[311] + 3(111)/M_x + (311)\,c_x - 3\,c_x^2/2 = 0$$
$$2M_x[331] + 2(311)/M_x + 2(331)\,c_x + 3c_x/M_x = 0$$
$$3M_x[333] + (331)/M_x + 3(333)\,c_x + 3(M_x^2 - 1)/2M_x^2 = 0$$
$$M_x[322] + (211)/M_x + (322)\,c_x - c_x^2/2 = 0$$
$$M_x[432] + (421)/M_x + (432)\,c_x + c_y/M_y = 0$$
$$M_x[443] + (441)/M_x + (443)\,c_x + (M_y^2 - 1)/2M_y^2 = 0$$
$$2M_x[221] - 2M_y[211] + 2(221)\,c_x - 2(211)\,c_y - M_y\,c_x^2\,c_y + \\ + M_x\,c_x\,c_y^2 + L\,c_x^2\,c_y^2 = 0$$
$$M_x[421] - 2M_y[411] + (421)\,c_x - 2\,(411)\,c_y + c_x^2 - M_x\,c_x\,c_y/M_y - \\ - L\,c_x^2\,c_y = 0$$
$$2M_x[322] - M_y[321] + 2(322)\,c_x - (321)\,c_y + M_y\,c_x\,c_y/M_x - c_y^2 - \\ - L\,c_x\,c_y^2/M_x = 0$$
$$M_x[432] - M_y[431] + (432)\,c_x - (431)\,c_y - c_x/M_x + c_y/M_y + \\ + L\,c_x\,c_y/M_x\,M_y = 0$$
$$M_x[421] + 2(211)/M_y + (421)\,c_x - M_x\,c_x\,c_y/M_y - L\,c_x^2 c_y/M_y = 0$$
$$2M_x[441] + 2(411)/M_y + 2(441)\,c_x + M_x\,c_x/M_y^2 + L\,c_x^2/M_y^2 = 0$$
$$M_x[432] + (321)/M_y + (432)\,c_x + c_y/M_y + L\,c_x\,c_y/M_x\,M_y = 0$$

* *Meads* comments that "These relationships are satisfied by the aberration coefficients calculated by the 7090 computer code in every instance checked. In most cases, the relations are valid to five or six significant figures."

$$
\begin{aligned}
&2M_x[443] + (431)/M_y + 2(443)\,c_x + 1 - 1/M_y^2 - L\,c_x/M_x\,M_y^2 = 0\\
&M_y[321] + 2(221)/M_x + (321)\,c_y - M_y\,c_x\,c_y - L\,c_x\,c_y^2/M_x = 0\\
&M_y[431] + (421)/M_x + (431)\,c_y + c_x/M_x - L\,c_x\,c_y/M_x\,M_y = 0\\
&2M_y[332] + 2(322)/M_x + 2(332)\,c_y + M_y\,c_y/M_x^2 - L\,c_y^2/M_x^2 = 0\\
&2M_y[433] + (432)/M_x + 2(433)\,c_y + 1 - 1/M_x^2 - L\,c_y/M_x^2\,M_y = 0\\
&(321)/M_y - (421)/M_x + L\,c_x\,c_y/M_x\,M_y = 0\\
&(431)/M_y - 2(441)/M_x - L\,c_x/M_x\,M_y^2 = 0\\
&2(332)/M_y - (432)/M_x - L\,c_y/M_x^2\,M_y = 0\\
&2(433)/M_y - 2(443)/M_x + L/M_x^2\,M_y^2 = 0\\
&M_y[411] + (211)/M_y + (411)\,c_y - c_x^2/2 = 0\\
&M_y[431] + (321)/M_y + (431)\,c_y + c_x/M_x = 0\\
&M_y[433] + (332)/M_y + (433)\,c_y + (M_x^2 - 1)/2M_x^2 = 0\\
&M_y[422] + 3(222)/M_y + (422)\,c_y - 3\,c_y^2/2 = 0\\
&2M_y[442] + 2(422)/M_y + 2(442)\,c_y + 3c_y/M_y = 0\\
&3M_y[444] + (442)/M_y + 3(444)\,c_y + 3(M_y^2 - 1)/2M_y^2 = 0\,. \qquad (3.55)
\end{aligned}
$$

In these equations, M_x and M_y are the "magnifications" in the x and y directions, and c_x, c_y are the convergences ($c_x = 1/f_x$, $c_y = 1/f_y$); in image space, we have, paraxially,

$$
\begin{pmatrix} x_c \\ x_c' \end{pmatrix} = \begin{pmatrix} \dfrac{z_{Fi} - z_c}{f_x} & f_x - \dfrac{(z_{Fi} - z_c)\,(z_o - z_{Fo})}{f_x} \\ -\dfrac{1}{f_x} & \dfrac{z_o - z_{Fo}}{f_x} \end{pmatrix} \begin{pmatrix} x_o \\ x_o' \end{pmatrix}
$$

so that if we denote the abscissae of the line images conjugate to z_o by z_x and z_y, then

$$
\begin{aligned}
M_x &= f_x/(z_o - z_{Fo}^{(x)}) = (z_{Fi}^{(x)} - z_x)/f_x\\
M_y &= f_y/(z_o - z_{Fo}^{(y)}) = (z_{Fi}^{(y)} - z_y)/f_y
\end{aligned}
$$

and

$$
\begin{aligned}
x_c &= \left(M_x + \frac{z_x - z_c}{f_x}\right) x_o + \frac{z_c - z_x}{M_x}\,x_o'\\
y_c &= \left(M_y + \frac{z_y - z_c}{f_y}\right) y_o + \frac{z_c - z_y}{M_y}\,y_o'\,.
\end{aligned}
$$

The quantities $[ijk]$ denote the aberration coefficients associated with the slopes of the emergent rays instead of with their positions:

$$
\begin{aligned}
x'(z) &= \psi_1\chi_5(z) + \psi_3\chi_7(z) + \sum_{1 \le k \le j \le i \le 6} [ijk]\,\psi_i\psi_j\psi_k\\
y'(z) &= \psi_2\chi_6(z) + \psi_4\chi_8(z) + \sum_{1 \le k \le j \le i \le 6} [ijk]\,\psi_i\psi_j\psi_k
\end{aligned}
$$

$[ijk]$ is obtained by replacing each χ_k by χ_{k+4} in (ijk), that is, by replacing each s, T by its derivative.

It is not difficult to rewrite this analysis in a form applicable to short quadrupoles, in which $\mathfrak{Q}$ is not separated into $\mathfrak{Q}_r$ and $\mathfrak{Q}_f$; we shall not pursue this, since formulae for such aberration coefficients have already been derived in the section on perturbation characteristics. In fact, we have only to replace $\mathfrak{Q}_r$ by $\mathfrak{Q}$ and x_r by x, and ignore $\mathfrak{Q}_f$ and x_f everywhere, to obtain the requisite formulae. This is regarded as disadvantageous by high energy beam opticians, because "one then loses the ready identification of the effects of the fringing field and also loses the calculational advantage of simple known first-order solutions" (*Meads*), but with the short quadrupoles used in microscopes and other optical instruments, the concept of a fringe field is misleadingly artificial; moreover, analytic solutions of the trajectory equations are available for potential models other than rectangular in shape.

3.4 Permissible aberrations and aberration patterns

In the preceding sections of this chapter on aberrations, we have shown how aberration coefficients, measures of the transverse shift of an electron beam when calculated to the paraxial and third-order approximations, may be deduced by a variety of mathematical techniques. In this section, we shall consider what types of shift are permissible in a twist-free system, and investigate the blemishes characteristic of each.

The position and slope of an electron ray in an arbitrary current plane are functions of four determining parameters and the distribution of refractive index. Common choices of the determining parameters are the position of the ray in two planes, normally the object and aperture planes, or the position and slope of the ray in one plane, the object plane, or the slope of the ray in the object and aperture planes. We select the first of these, so that

$$x(z_c) = x(x_o, y_o, x_a, y_a)$$
$$y(z_c) = y(x_o, y_o, x_a, y_a)\,.$$

Expanding these functions as power series, we obtain

$$x_c = (1000)_x\, x_o + (0100)_x\, y_o + (0010)_x\, x_a + (0001)_x\, y_a + \\ + \sum_{\alpha,\beta,\gamma,\delta} (\alpha\,\beta\,\gamma\,\delta)_x\, x_o^\alpha\, y_o^\beta\, x_a^\gamma\, y_a^\delta \quad (\alpha+\beta+\gamma+\delta=3)\,.$$

Since the system is twist-free, x_c must be transformed into $-x_c$ if x_o and x_a are replaced by $-x_o$ and $-x_a$ respectively. Thus all the coefficients $(\alpha\,\beta\,\gamma\,\delta)_x$ for which $\beta+\delta$ is odd vanish, and all the $(\alpha\,\beta\,\gamma\,\delta)_y$ for which $\alpha+\gamma$ is odd also vanish. Dropping the suffices x, y, we find

$$\begin{aligned} x = {} & g_x\, x_o + h_x\, x_a + \\ & + \{(3000)\, x_o^2 + (1200)\, y_o^2\}\, x_o + \{(0030)\, x_a^2 + (0012)\, y_a^2\}\, x_a \\ & + \{(1020)\, x_a^2 + (1002)\, y_a^2\}\, x_o + \{(2010)\, x_o^2 + (0210)\, y_o^2\}\, x_a + \\ & + \{(1101)\, x_o + (0111)\, x_a\}\, y_o\, y_a\,. \end{aligned} \tag{3.56a}$$

$$\begin{aligned} y = {} & g_y\, y_o + h_y\, y_a + \\ & + \{(2100)\, x_o^2 + (0300)\, y_o^2\}\, y_o + \{(0021)\, x_a^2 + (0003)\, y_a^2\}\, y_a + \\ & + \{(0120)\, x_a^2 + (0102)\, y_a^2\}\, y_o + \{(2001)\, x_o^2 + (0201)\, y_o^2\}\, y_a + \\ & + \{(1110)\, y_o + (1011)\, y_a\}\, x_o\, x_a\,. \end{aligned} \tag{3.56b}$$

To reveal any inter-relations between these coefficients ($\alpha\ \beta\ \gamma\ \delta$), we write the point perturbation characteristic function in terms of the invariants of the system:

$$V^{(4)} = \begin{pmatrix} x_o^2 \\ y_o^2 \\ x_a^2 \\ y_a^2 \\ x_o x_a \end{pmatrix}' \begin{pmatrix} 4000 & 2200 & 2020 & 2002 & 3010 & 2101 \\ 0 & 0400 & 0220 & 0202 & 1210 & 0301 \\ 0 & 0 & 0040 & 0022 & 1030 & 0121 \\ 0 & 0 & 0 & 0004 & 1012 & 0103 \\ 0 & 0 & 0 & 0 & 0 & 1111 \end{pmatrix} \begin{pmatrix} x_o^2 \\ y_o^2 \\ x_a^2 \\ y_a^2 \\ x_o x_a \\ y_o y_a \end{pmatrix}$$

and use equations (3.21) to obtain

$$\left.\begin{aligned} k_x(3000) &= 4h_{xc}[4000] - g_{xc}(3010) \\ k_x(1200) &= 2h_{xc}[2200] - g_{xc}(1210) \\ k_y(0300) &= 4h_{yc}[0400] - g_{yc}(0301) \\ k_y(2100) &= 2h_{yc}[2200] - g_{yc}(2101) \end{aligned}\right\}\text{(distortions)}$$

$$\left.\begin{aligned} k_x(2010) &= 3h_{xc}[3010] - 2g_{xc}(2020) \\ k_x(0210) &= h_{xc}[1210] - 2g_{xc}(0220) \\ k_x(1101) &= 2h_{xc}[2101] - g_{xc}(1111) \\ k_y(2001) &= h_{yc}[2101] - 2g_{yc}(2002) \\ k_y(0201) &= 3h_{yc}[0301] - 2g_{yc}(0202) \\ k_y(1110) &= 2h_{yc}[1210] - g_{yc}(1111) \end{aligned}\right\}\text{(astigmatisms)}$$

$$\left.\begin{aligned} k_x(1020) &= 2h_{xc}[2020] - 3g_{xc}(1030) \\ k_x(1002) &= 2h_{xc}[2002] - g_{xc}(1012) \\ k_x(0111) &= h_{xc}[1111] - 2g_{xc}(0121) \\ k_y(0120) &= 2h_{yc}[0220] - g_{yc}(0121) \\ k_y(0102) &= 2h_{yc}[0202] - 3g_{yc}(0103) \\ k_y(1011) &= h_{yc}[1111] - 2g_{yc}(1012) \end{aligned}\right\}\text{(comas)}$$

$$\left.\begin{aligned} k_x(0030) &= h_{xc}[1030] - 4g_{xc}(0040) \\ k_x(0012) &= h_{xc}[1012] - 2g_{xc}(0022) \\ k_y(0003) &= h_{yc}[0103] - 4g_{yc}(0004) \\ k_y(0021) &= h_{yc}[0121] - 2g_{yc}(0022) \end{aligned}\right\}\text{(aperture aberrations)}\,. \tag{3.57}$$

$[p\ q\ r\ s]$ denotes integration from z_a to z_c and $(p\ q\ r\ s)$ from z_o to z_c.

The usual way of displaying the effects of the various aberrations, other than the distortions, is to consider the locus of the points of intersection with the current plane of rays emerging from a fixed object point and intersecting the aperture plane along some simple curve, commonly a circle (*Chako* [27, 28], *Burfoot* [23, 24], *Amboss* [1], *Hawkes* [85, 86, 87, 96]) or an ellipse (*Meads* [130]). It can be shown very simply that the astigmatisms produce an ellipse, centred on the point ($g_{xc}\, x_o$, $g_{yc}\, y_o$), but the total effect of the comas is complex; further details and several possible coma patterns are to be found in *Burfoot* [23, 24] and in *Hawkes* [96], where the relationship between *Burfoot's* coefficients and those employed above is explained. Here, we consider in detail only the aperture aberrations and, to a lesser extent, the distortions. Some general discussion of symmetry and matrix elements is to be found in [61a].

The aperture aberrations

Writing ($\gamma\ \delta$) for ($\alpha\ \beta\ \gamma\ \delta$), we have

$$x_c = x_o\, g_{xc} + x_a\{h_{xc} + (30)\, x_a^2 + (12)\, y_a^2\}$$

$$y_c = y_o\, g_{yc} + y_a\{h_{yc} + (03)\, y_a^2 + (21)\, x_a^2\}$$

and it is clear that, in the image plane ($z = z_i$) of a stigmatic system (where $h_{xi} = h_{yi} = 0$) which is also orthomorphic ($g_{xi}/k_x = 1/h'_{xi} = g_{yi}/k_y = 1/h'_{yi}$), (12) = (21). (If we had expressed the aberration coefficients in terms of the angle at which the rays intersect the image plane, so that

$$x_c - x_o\, g_{xi} = (30)^*\, \alpha^3 + (12)^*\, \alpha\, \beta^2$$

$$y_c - y_o\, g_{yi} = (03)^*\, \beta^3 + (21)^*\, \alpha^2\, \beta$$

with $\alpha = x_a\, h'_{xi}$ and $\beta = y_a\, h'_{yi}$, we should have found that the coefficients (12)* and (21)* were always equal at z_i, even for an anamorphotic system [39].) If we consider rays which intersect the aperture plane around the periphery of an ellipse, we write

$$x_a = \xi_a \cos\theta \quad y_a = \eta_a \sin\theta$$

giving

$$x_c - x_o\, g_{xc} = \xi_a [\{h_{xc} + (30)\, \xi_a^2\} \cos\theta + \{(12)\, \eta_a^2 - (30)\, \xi_a^2\} \cos\theta \sin^2\theta]$$

$$y_c - y_o\, g_{yc} = \eta_a [\{h_{yc} + (03)\, \eta_a^2\} \sin\theta + \{(21)\, \xi_a^2 - (03)\, \eta_a^2\} \cos^2\theta \sin\theta]$$

or writing

$$x_c - x_o\, g_{xc} = (\varkappa + \lambda \sin^2\theta) \cos\theta$$

$$y_c - y_o\, g_{yc} = (\mu + \nu \cos^2\theta) \sin\theta$$

we obtain [130]

$$X = \{(1 + \varepsilon) + \alpha \sin^2\theta\} \cos\theta$$

$$Y = \{(1 - \varepsilon) + \alpha \cos^2\theta\} \sin\theta$$

with

$$X = \frac{2\nu}{\varkappa\nu + \lambda\mu}\, x \qquad Y = \frac{2\lambda}{\varkappa\nu + \lambda\mu}\, y$$

$$\varepsilon = \frac{\varkappa\nu - \lambda\mu}{\varkappa\nu + \lambda\mu} \qquad \alpha = \frac{2\lambda\nu}{\varkappa\nu + \lambda\mu}\,.$$

The parameter α is small except in the neighbourhood of the line images, since the numerator contains only aberration terms, while the denominator involves h_x and h_y; unless the astigmatic difference is small or zero, α will usually be small everywhere.

Meads classifies the shape of the aberration figures according to the values of ε and α. (Only positive values of ε need be considered.) The figure will have no loops if the only solutions of $X(\theta) = 0$ occur at $\theta = n\pi/2$ and if the only solutions of $Y(\theta) = 0$ occur for $\theta = n\pi$; this will be the case if $(1+\varepsilon)/(-\alpha) < 0$ and $(1-\varepsilon)/(-\alpha) < 0$. The first condition is always true for $\alpha > 0$ and the second is true for $\alpha > 0$ if $\varepsilon < 1$. The first is never satisfied for $\alpha < 0$ and the second is satisfied for $\alpha < 0$ if $\varepsilon > 1$.

If $\alpha < 0$, $X(\theta)$ has a solution provided that $1 + \varepsilon + \alpha < 0$; if $\alpha > 0$, $Y(\theta)$ has a solution provided that $\varepsilon > 1$ and $1 + \alpha - \varepsilon < 0$, and if $\alpha < 0$, $Y(\theta)$ has a solution if $\varepsilon < 1$ and again $1 + \alpha - \varepsilon < 0$. The same value of θ gives $X(\theta) = Y(\theta) = 0$ if $\alpha = -2$ and $0 < \varepsilon < 1$.

The figures are concave at $Y = 0$ for $2\alpha > \varepsilon + 1$ and at $X = 0$ for $2\alpha > 1 - \varepsilon$. The polar form of the curves, $X^2(\theta) + Y^2(\theta) = R^2(\theta) = 0$, has four maxima and four minima in the zone $|4\varepsilon| \leqq |(\alpha+2)^2 - 4|$ and two maxima and two minima outside this region; the dividing curves are parabolae. When α lies between -4 and 0, $R^2 \leqq (1+\varepsilon)^2$, and outside these limits $R^2 \geqq (1-\varepsilon)^2$. All these features of the aberration figures are illustrated in Fig. 6a, and in Fig. 6b—d, aberration figures for a wide range of values of α and ε are shown.

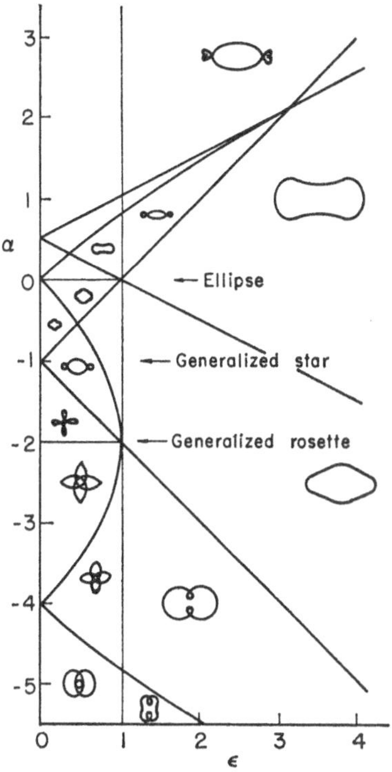

Fig. 6a. The division of the $\alpha - \varepsilon$ plane into regions producing different aberration patterns

The aperture aberrations of stigmatic quadrupole systems are commonly classified into spherical aberration, "star" and "rosette" aberration (*Burfoot* [23, 24]). We write $\xi_a = \eta_a = r_a$, and

$$x_i - x_o g_{xi} = x_a\{\alpha_x r_a^2 + \beta x_a^2 + \gamma(x_a^2 - y_a^2)\}$$

$$y_i - y_o g_{yi} = y_a\{\alpha_y r_a^2 - \beta y_a^2 - \gamma(x_a^2 - y_a^2)\}$$

so that

$$\alpha_x = \frac{1}{4}(30) + \frac{1}{4}(03) + \frac{3}{4}(12) - \frac{1}{4}(21)$$

$$\alpha_y = \frac{1}{4}(30) + \frac{1}{4}(03) - \frac{1}{4}(12) + \frac{3}{4}(21)$$

$$\beta = \frac{1}{2}\{(30) - (03) - (12) + (21)\}$$

$$\gamma = \frac{1}{4}\{(30) + (03) - (12) - (21)\}\,.$$

With $x_a^2 + y_a^2 = r_a^2 =$ constant, α_x and α_y produce an ellipse in the image plane which degenerates into a circle, the spherical aberration of round lenses, if the system is orthomorphic [(12) = (21)]. The coefficient β is a measure of the star aberration, and γ of the rosette.

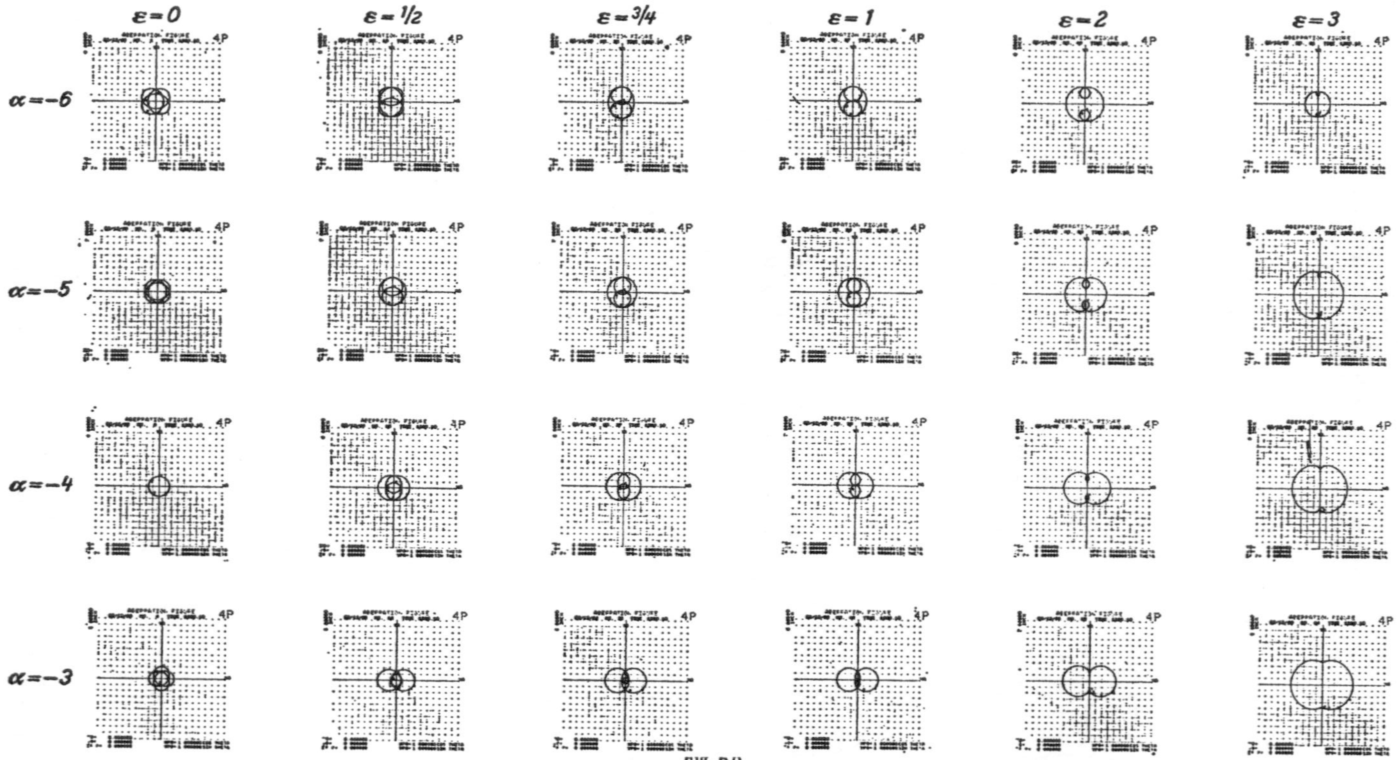

Fig. 6b

Fig. 6b—d. Typical aberration figures, $0 \leqq \varepsilon \leqq 3$; $-6 \leqq \alpha \leqq 2$. (By courtesy of Dr. *Meads* and the University of California Lawrence Radiation Laboratory, Berkeley)

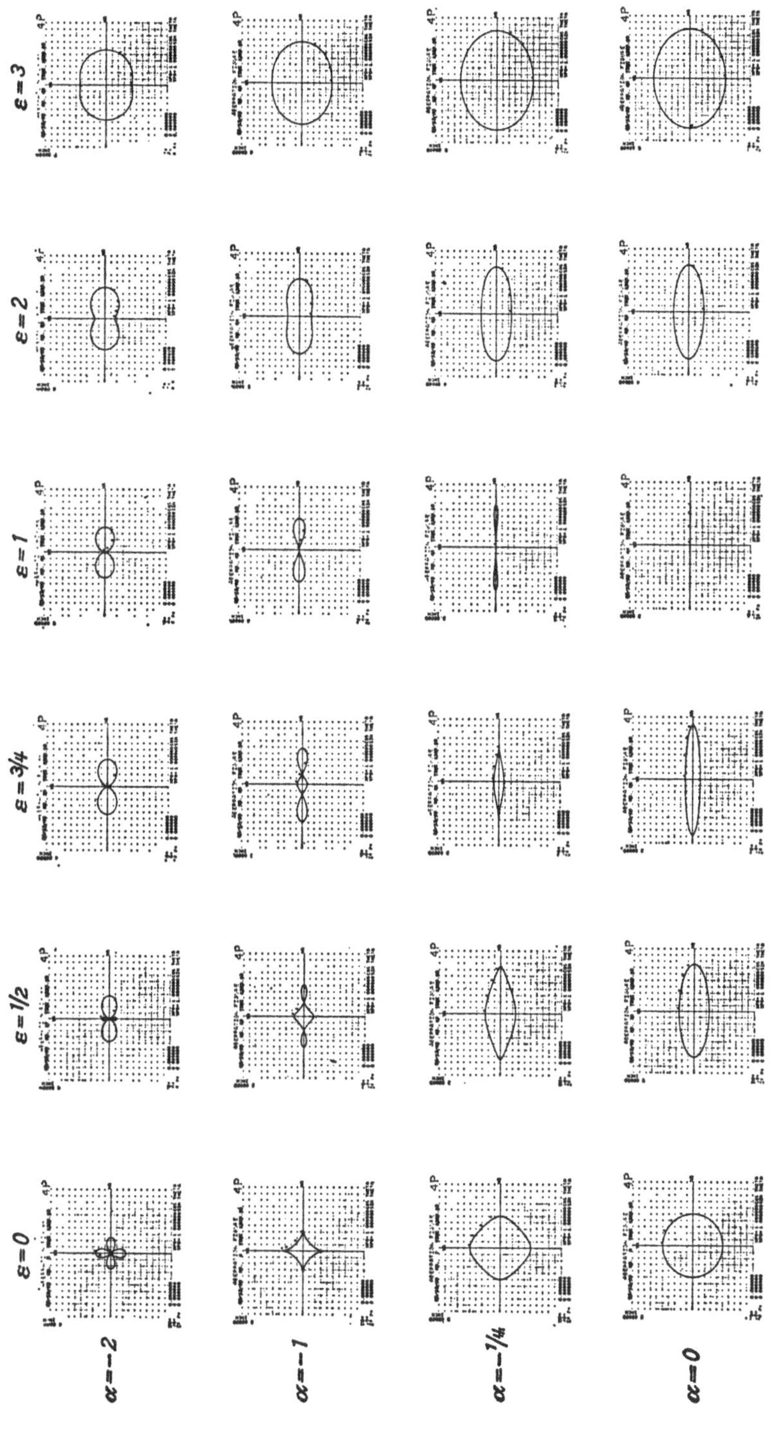

Fig. 6c

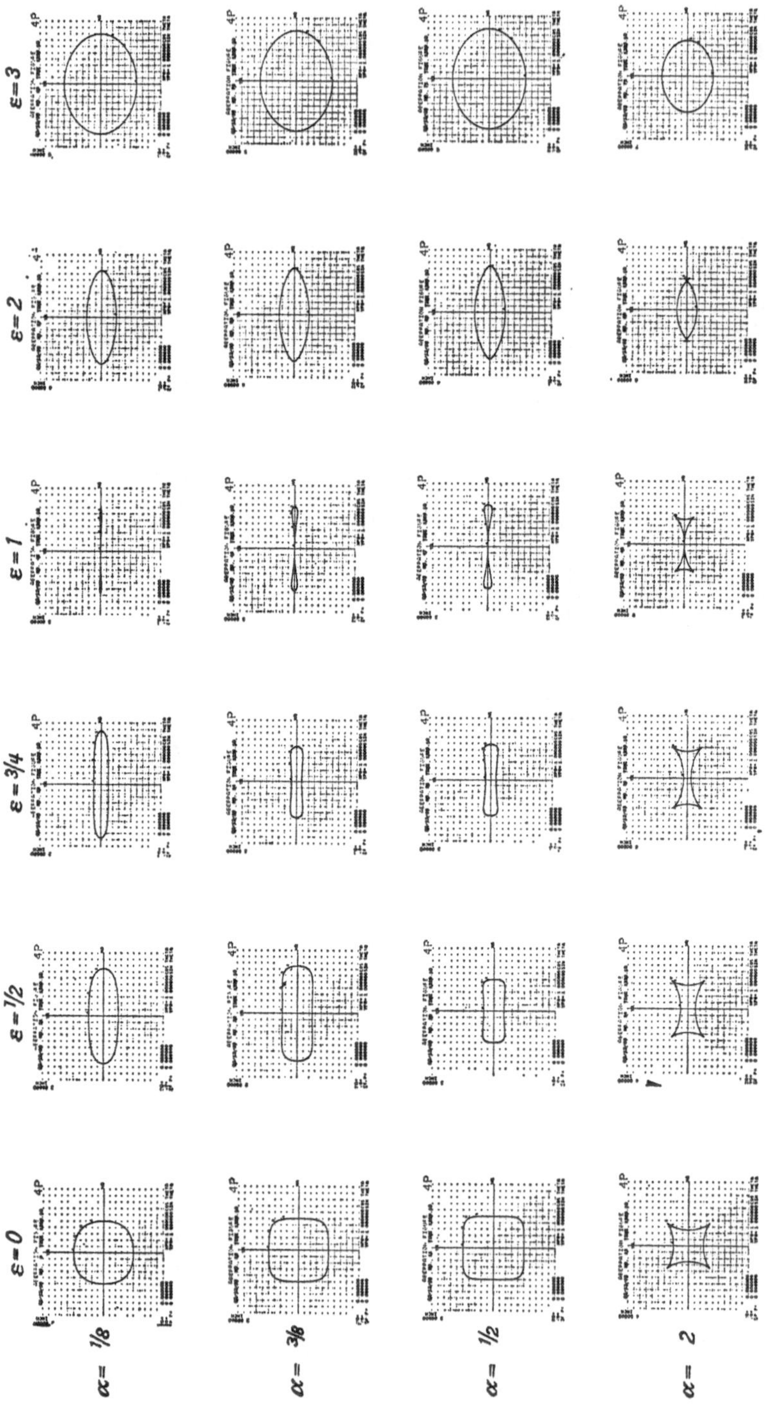

Fig. 6d

The distortions

We now use $(\alpha\,\beta)$ to denote $(\alpha\,\beta\,0\,0)$, so that

$$x_c - h_{xc}\,x_a = x_o\{g_{xc} + (30)\,x_o^2 + (12)\,y_o^2\}$$
$$y_c - h_{yc}\,y_a = y_o\{g_{yc} + (03)\,y_o^2 + (21)\,x_o^2\}\,.$$

The distortions of a rectangular grid are considered by *Burfoot* ([23, 24], cf. *Hawkes* [96]); if we consider circles in the object plane, $r_o =$ constant, we can write

$$x_c - h_{xc}\,x_a = \{g_{xc} + r_o^2(\varkappa + \mu\cos 2\theta_o)\}\,x_o$$
$$y_c - h_{yc}\,y_a = \{g_{yc} + r_o^2(\lambda + \nu\cos 2\theta_o)\}\,y_o\,.$$

Thus $\varkappa$ and λ, acting alone, would convert the circle $r_o =$ constant into an ellipse; μ and ν have much the same effect as aperture aberrations upon an astigmatic beam.

3.5. Mechanical aberrations

As we mentioned at the beginning of § 3, first-order perturbation theory yields the effects of any small change from $m^{(2)}$ to $m^{(2)} + m^{(P)}$. We have dealt in considerable detail with the geometrical aberrations, and we postpone further discussion of the chromatic aberrations to § 5; to conclude this section, therefore, we briefly examine the mechanical aberrations of quadrupoles, and the tolerances which will have to be achieved if these defects are not to be a problem.

No study of the mechanical aberrations of quadrupoles has yet been made, comparable in thoroughness with *Bertein's* and *Sturrock's* analyses of round electrostatic and magnetic lenses. Here, therefore, a separate account is given of each of the publications dealing with the problem, as an attempt to make a general synthesis seems premature.

a) Das Auflösungsvermögen sphärisch korrigierter elektrostatischer Elektronenmikroskope by W. E. Meyer [134, 135]. Whereas the potential distribution in a perfect system of quadrupoles, octopoles and round lenses is given by

$$\varphi_{id} = \Phi - \frac{1}{4}(\Phi'' - D)\,x^2 - \frac{1}{4}(\Phi'' + D)\,y^2 +$$
$$+ \left(\frac{1}{64}\Phi^{(\mathrm{iv})} - \frac{1}{48}D'' + D_1\right)x^4 + \left(\frac{1}{64}\Phi^{(\mathrm{iv})} + \frac{1}{48}D'' + D_1\right)y^4 +$$
$$+ \left(\frac{1}{32}\Phi^{(\mathrm{iv})} - 6D_1\right)x^2y^2\,,$$

the potential in a faulty system will contain perturbation terms in addition to the "ideal" φ_{id}; the perturbing potential φ_{st} *(Störpotential)* is of the form

$$\varphi_{st} = \sum_n \varphi_n$$

with

$$\varphi_0 = \Phi_\Delta; \quad \varphi_1 = \Phi_{1c}\, x + \Phi_{1s}\, y;$$

$$\varphi_2 = \frac{1}{4}\,(D_c - \Phi_\Delta'')\, x^2 - \frac{1}{4}\,(D_c + \Phi_\Delta'')\, y^2 + P\, x\, y;$$

$$\varphi_3 = \left(\Phi_{3c} - \frac{1}{8}\,\Phi_{1c}''\right) x^3 - \left(\Phi_{3s} + \frac{1}{8}\,\Phi_{1s}''\right) y^3 + \\ + \left(3\Phi_{3s} - \frac{1}{8}\,\Phi_{1s}''\right) x^2 y - \left(3\Phi_{3c} + \frac{1}{8}\,\Phi_{1c}''\right) x\, y^2;$$

$$\varphi_4 = \left(\frac{1}{64}\,\Phi_\Delta^{(\mathrm{iv})} - \frac{1}{48}\,D_c'' + D_{1c}\right) x^4 + \left(\frac{1}{64}\,\Phi_\Delta^{(\mathrm{iv})} + \frac{1}{48}\,D_c'' + D_{1c}\right) y^4 + \\ + \left(4P_1 - \frac{1}{12}\,P''\right) x^3 y - \left(4P_1 + \frac{1}{12}\,P''\right) x\, y^2 + \\ + \left(\frac{1}{32}\,\Phi_\Delta^{(\mathrm{iv})} - 6D_{1c}\right) x^2 y^2 .$$

Elements which would produce each of the potentials* Φ_{1c}, Φ_{1s}, D_c, P, Φ_{3c}, Φ_{3s}, D_{1c} and P_1 are shown in Fig. 7. The equations of motion for an electron travelling through a potential $\varphi_{id} + \varphi_{st}$ are now of the form

$$\Phi x'' + \frac{1}{2}\,\Phi' x' + \frac{1}{4}\,(\Phi'' - D)\, x = R_x + J_x$$

$$\Phi y'' + \frac{1}{2}\,\Phi' y' + \frac{1}{4}\,(\Phi'' + D)\, y = R_y + J_y$$

in which R_x and R_y represent the terms arising from φ_{id}, and J_x, J_y, the parasitic terms created by φ_{st}.

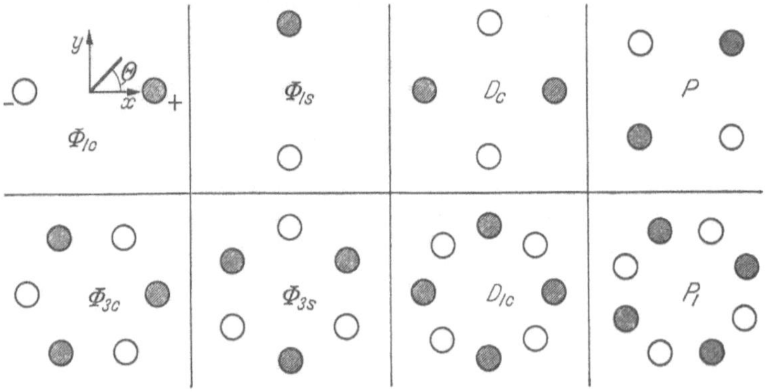

Fig. 7. The elements producing each of the types of parasitic potential in an electrostatic system. (By courtesy of Dr. *Meyer* in *Optik* [135])

Setting $\xi = x_i/M_x$ and $\eta = y_i/M_y$, *Meyer* finds that in the stigmatic image plane, the parasitic aberrations of each order are as follows** $(U = \sqrt{\Phi})$:

* In *Meyer's* notation, $D_c = 4\Phi_{2c}$, $D_{1c} = \Phi_{4c}$, $P = 2\Phi_{2s}$, $P_1 = \Phi_{4s}$, $D = 4\Phi_2$ and $D_1 = \Phi_4$.

** We recall that $T_{x0} = T_{y0} = 0$, $T_{x0}' = T_{x0}' = 1$; *Meyer* writes x_α for T_x and y_α for T_y.

Zero order

$$\xi = \frac{1}{2U_o}\int\limits_o^i \frac{\Phi_{1c}}{U} T_x \,\mathrm{d}z\,; \quad \eta = \frac{1}{2U_o}\int\limits_o^i \frac{\Phi_{1s}}{U} T_y \,\mathrm{d}z\,.$$

ξ and η do not vary with x_o' and y_o', and are hence harmless provided that φ_1 does not vary with time. (Time-dependent perturbations are considered in [136].)

First order

$$\xi = \frac{x_o'}{U_o}\int\limits_o^i \frac{\delta + D_c/4}{U} T_x^2 \,\mathrm{d}z + \frac{y_o'}{U_o}\int\limits_o^i \frac{P}{2U} T_x T_y \,\mathrm{d}z$$

$$\eta = \frac{y_o'}{U_o}\int\limits_o^i \frac{\delta - D_c/4}{U} T_y^2 \,\mathrm{d}z + \frac{x_o'}{U_o}\int\limits_o^i \frac{P}{2U} T_x T_y \,\mathrm{d}z$$

in which δ is a measure of the defocusing, so that the additional effect is an astigmatism with which a stigmator can cope.

Second order

$$\xi = A_3 x_o'^2 + A_4 x_o' y_o' + A_5 y_o'^2$$
$$\eta = B_3 y_o'^3 + B_4 x_o' y_o' + B_5 x_o'^2$$

in which

$$A_3 = \frac{1}{U_o}\int\limits_o^i U\left(\frac{3}{2}\frac{\Phi_{3c}}{\Phi} + \frac{3}{8}\frac{\Phi_{1c}\Phi'^2}{\Phi^3} - \frac{\Phi_{1c} D}{4\Phi^2} - \frac{11}{32}\frac{\Phi_{1c}'\Phi'}{\Phi^2} + \right.$$
$$\left. + \frac{1}{16}\frac{\Phi_{1c}'}{\Phi}\frac{T_x'}{T_x} - \frac{\Phi_{1c}\Phi'}{4\Phi^2}\frac{T_x'}{T_x} - \frac{\Phi_{1c}}{2\Phi}\frac{T_x'^2}{T_x^2}\right) T_x^3 \,\mathrm{d}z$$

$$A_4 = \frac{1}{U_o}\int\limits_o^i U\left(3\frac{\Phi_{3s}}{\Phi} + \frac{3}{8}\frac{\Phi_{1s}\Phi'^2}{\Phi^3} - \frac{\Phi_{1s} D}{4\Phi^2} - \frac{5}{16}\frac{\Phi_{1s}'\Phi'}{\Phi^2} - \frac{\Phi_{1s}'}{4\Phi}\frac{T_x'}{T_x} - \right.$$
$$\left. - \frac{\Phi_{1s}\Phi'}{4\Phi^2}\frac{T_y'}{T_y} + \frac{\Phi_{1s}'}{8\Phi}\frac{T_y'}{T_y} - \frac{\Phi_{1s}}{2\Phi}\frac{T_x' T_y'}{T_x T_y}\right) T_x^2 T_y \,\mathrm{d}z$$

$$A_5 = \frac{1}{U_0}\int\limits_o^i U\left(-\frac{3}{2}\frac{\Phi_{3c}}{\Phi} - \frac{\Phi_{1c}'\Phi'}{32\Phi^2} + \frac{\Phi_{1c}'}{16\Phi}\frac{T_x'}{T_x} + \frac{\Phi_{1c}'}{8\Phi}\frac{T_y'}{T_y}\right) T_x T_y^2 \,\mathrm{d}z$$

$$B_3 = \frac{1}{U_o}\int\limits_o^i U\left(-\frac{3}{2}\frac{\Phi_{3s}}{\Phi} + \frac{3}{8}\frac{\Phi_{1s}\Phi'^2}{\Phi^3} + \frac{\Phi_{1c} D}{4\Phi^2} - \frac{11}{32}\frac{\Phi_{1s}'\Phi'}{\Phi^2} + \right.$$
$$\left. + \frac{\Phi_{1s}'}{16\Phi}\frac{T_y'}{T_y} - \frac{\Phi_{1s}\Phi'}{4\Phi^2}\frac{T_y'}{T_y} - \frac{\Phi_{1s}}{2\Phi}\frac{T_y'^2}{T_y^2}\right) T_y^3 \,\mathrm{d}z$$

$$B_4 = \frac{1}{U_o}\int\limits_o^i U\left(-3\frac{\Phi_{3c}}{\Phi} + \frac{3}{8}\frac{\Phi_{1c}\Phi'^2}{\Phi^3} + \frac{\Phi_{1c} D}{4\Phi^2} - \frac{5}{16}\frac{\Phi_{1c}'\Phi'}{\Phi^2} - \right.$$
$$\left. - \frac{\Phi_{1c}'}{4\Phi}\frac{T_y'}{T_y} - \frac{\Phi_{1c}\Phi'}{4\Phi^2}\frac{T_x'}{T_x} + \frac{\Phi_{1c}'}{8\Phi}\frac{T_x'}{T_x} - \frac{\Phi_{1c}}{2\Phi}\frac{T_x' T_y'}{T_x T_y}\right) T_x T_y^2 \,\mathrm{d}z$$

$$B_5 = \frac{1}{U_o}\int\limits_o^i U\left(\frac{3}{2}\frac{\Phi_{3s}}{\Phi} - \frac{\Phi_{1s}'\Phi'}{32\Phi^2}\ \frac{\Phi_{1s}'}{16\Phi}\frac{T_y'}{T_y} + \frac{\Phi_{1s}'}{8\Phi}\frac{T_x'}{T_x}\right) T_x^2 T_y \,\mathrm{d}z\,.$$

Setting $x_o'^2 + y_o'^2 = \omega^2$ and $x_o' = \omega\cos\Theta$, $y_o' = \omega\sin\Theta$, and writing $a_i = \omega^2 A_i$, we find

$$\begin{aligned}\xi &= a_3\cos^2\Theta + a_4\sin\Theta\cos\Theta + a_5\sin^2\Theta\\ &= \frac{a_3+a_5}{2} + \frac{a_3-a_5}{2}\cos 2\Theta + \frac{a_4}{2}\sin 2\Theta\\ \eta &= b_3\sin^2\Theta + b_4\sin\Theta\cos\Theta + b_5\cos^2\Theta\\ &= \frac{b_3+b_5}{2} - \frac{b_3-b_5}{2}\cos 2\Theta + \frac{b_4}{2}\sin 2\Theta\end{aligned}$$

which represents an ellipse centred on the point $(a_3 + a_5)/2$, $(b_3 + b_5)/2$; if $\Phi_{1c} = \Phi_{1s} = 0$, $a_4 = 2b_5$ and $b_4 = 2a_5$, and if in addition the imaging system contains only round lenses, $a_5 = -a_3$ and $b_5 = -b_3$ so that the aberration figure becomes a circle, centred on the axis, radius $\sqrt{a_3^2 + b_3^2}$. Outside the image plane, however, a triangular figure is produced. A sextupole cannot correct this aberration completely, because of the relations $a_4 = 2b_5$ and $b_4 = 2a_5$, and a "Φ_1-stigmator" (dipole) is necessary; the latter must be situated within the lenses or at points where the slopes of the rays do not vanish.

Third order

$$\begin{aligned}\xi &= A_6 x_o'^3 + A_7 x_o'^2 y_o' + A_8 x_o' y_o'^2 + A_9 y_o'^3\\ \eta &= B_6 y_o'^3 + B_7 x_o' y_o'^2 + B_8 x_o'^2 y_o' + B_9 x_o'^3\end{aligned}$$

in which

$$A_6 = \frac{1}{U_o}\int_o^i U\left(\frac{E}{\Phi} + 2\frac{D_{1c}}{\Phi} + \frac{3}{32}\frac{D_c\Phi'^2}{\Phi^3} - \frac{D_c D}{16\Phi^2} - \frac{D_c'\Phi'}{12\Phi^2} - \right.$$
$$\left. - \frac{D_c\Phi'}{8\Phi^2}\frac{T_x'}{T_x} + \frac{D_c'}{24\Phi}\frac{T_x'}{T_x} - \frac{D_c}{4\Phi}\frac{T_x'^2}{T_x^2}\right) T_x^4\,\mathrm{d}z$$

$$A_7 = \frac{1}{U_o}\int_o^i U\left\{6\frac{P_1}{\Phi} + \frac{3}{8}\frac{P\Phi'^2}{\Phi^3} - \frac{PD}{4\Phi^2} - \frac{5}{16}\frac{P'\Phi'}{\Phi^2} - \frac{P\Phi'}{4\Phi^2}\times\right.$$
$$\left.\times\left(\frac{T_x'}{T_x} + \frac{T_y'}{T_y}\right) + \frac{P'}{8\Phi}\left(\frac{T_y'}{T_y} - \frac{T_x'}{T_x}\right) - \frac{P}{2\Phi}\frac{T_x'}{T_x}\left(\frac{T_x'}{T_x} + \frac{T_y'}{T_y}\right)\right\} T_x^3 T_y\,\mathrm{d}z$$

$$A_8 = \frac{1}{U_o}\int_o^i U\left(\frac{G}{\Phi} - \frac{6D_{1c}}{\Phi} - \frac{3}{32}\frac{D_c\Phi'^2}{\Phi^3} + \frac{D_c D}{16\Phi^2} + \frac{D_c'\Phi'}{16\Phi^2} + \right.$$
$$\left. + \frac{D_c'}{8\Phi}\frac{T_x'}{T_x} + \frac{D_c\Phi'}{8\Phi^2}\frac{T_y'}{T_y} + \frac{D_c}{4\Phi}\frac{T_x'T_y'}{T_x T_y}\right) T_x^2 T_y^2\,\mathrm{d}z$$

$$A_9 = \frac{1}{U_o}\int_o^i U\left(-2\frac{P_1}{\Phi} - \frac{P'\Phi'}{48\Phi^2} + \frac{P'}{24\Phi}\frac{T_x'}{T_x} + \frac{P'}{8\Phi}\frac{T_y'}{T_y}\right) T_x T_y^3\,\mathrm{d}z$$

$$B_6 = \frac{1}{U_o}\int_o^i U\left(\frac{F}{\Phi} + 2\frac{D_{1c}}{\Phi} - \frac{3}{32}\frac{D_c\Phi'^2}{\Phi^3} - \frac{D_c D}{16\Phi^2} + \frac{D_c'\Phi'}{12\Phi^2} + \right.$$
$$\left. + \frac{D_c\Phi'}{8\Phi^2}\frac{T_y'}{T_y} - \frac{D_c'}{24\Phi}\frac{T_y'}{T_y} + \frac{D_c}{4\Phi}\frac{T_y'^2}{T_y^2}\right) T_y^4\,\mathrm{d}z$$

$$B_7 = \frac{1}{U_o} \int_o^i U \left\{ -6 \frac{P_1}{\Phi} + \frac{3}{8} \frac{P \Phi'^2}{\Phi^3} + \frac{PD}{4\Phi^2} - \frac{5}{16} \frac{P' \Phi'}{\Phi^2} - \frac{P \Phi'}{4\Phi^2} \times \right.$$

$$\left. \times \left(\frac{T'_x}{T_x} + \frac{T'_y}{T_y} \right) + \frac{P'}{8\Phi} \left(\frac{T'_x}{T_x} - \frac{T'_y}{T_y} \right) - \frac{P}{2\Phi} \frac{T'_y}{T_y} \left(\frac{T'_x}{T_x} + \frac{T'_y}{T_y} \right) \right\} T_x T_y^3 \, \mathrm{d}z$$

$$B_8 = \frac{1}{U_o} \int_o^i U \left(\frac{G}{\Phi} - 6 \frac{D_{1c}}{\Phi} + \frac{3}{32} \frac{D_c \Phi'^2}{\Phi^3} + \frac{D_c D}{16\Phi^2} - \frac{D'_c \Phi'}{16\Phi^2} - \right.$$

$$\left. - \frac{D_c}{8\Phi} \frac{T'_y}{T_y} - \frac{D_c \Phi'}{8\Phi^2} \frac{T'_x}{T_x} - \frac{D_c}{4\Phi} \frac{T'_x T'_y}{T_x T_y} \right) T_x^2 T_y^2 \, \mathrm{d}z$$

$$B_9 = \frac{1}{U_o} \int_o^i U \left(2 \frac{P_1}{\Phi} - \frac{P' \Phi'}{48\Phi^2} + \frac{P'}{24\Phi} \frac{T'_y}{T_y} + \frac{P'}{8\Phi} \frac{T'_x}{T_x} \right) T_x^3 T_y \, \mathrm{d}z \, .$$

These aberrations may be written

$$\xi = a_6 \cos^3\Theta + a_7 \cos^2\Theta \sin\Theta + a_8 \cos\Theta \sin^2\Theta + a_9 \sin^3\Theta$$

$$\eta = b_6 \sin^3\Theta + b_7 \cos\Theta \sin^2\Theta + b_8 \cos^2\Theta \sin\Theta + b_9 \cos^3\Theta$$

or

$$\xi = \frac{a_6 - a_8}{4} \cos 3\Theta + \frac{3a_6 + a_8}{4} \cos\Theta + \frac{a_7 - a_9}{4} \sin 3\Theta + \frac{a_7 + 3a_9}{4} \sin\Theta$$

$$\eta = \frac{b_8 - b_6}{4} \sin 3\Theta + \frac{3b_6 + b_8}{4} \sin\Theta + \frac{b_9 - b_7}{4} \cos 3\Theta + \frac{b_7 + 3b_9}{4} \cos\Theta \, .$$

Again, all the aberrations cannot be eliminated with the aid of octopoles, for if $D_c = P = 0$, we see that $b_7 = 3a_9$, $a_8 = b_8$ and $a_7 = 3b_9$; quadrupole stigmators will also be necessary, and these can be designed to eliminate both third and first order effects.

The remainder of *Meyer's* paper is concerned with the delicate problem of adjusting a corrected microscope in practice, and devising tests sensitive enough to check the adjustment. The relationship between the functions appearing in φ_{st} and the mechanical defects of the electrodes remains unresolved, however.

b) The work of J. H. M. Deltrap on magnetic quadrupoles [39]. *Deltrap* first relates the functions G_1, H, Δ_1 and Q_1 which appear in the magnetostatic potential φ^m to the imperfections in the pole-pieces of quadrupoles producing φ^m; following *Glaser* [74],

$$\varphi^m = \Phi_m - \frac{1}{4} \Phi''_m r^2 + \frac{1}{64} \Phi_m^{(\mathrm{iv})} r^4 +$$

$$+ \cos 2\theta \left(\frac{1}{4} \Delta r^2 - \frac{1}{48} \Delta'' r^4 \right) + \sin 2\theta \left(\frac{1}{2} Q r^2 - \frac{1}{24} Q'' r^4 \right) +$$

$$+ \frac{1}{3} G_1 r^3 \cos 3\theta - \frac{1}{3} H_1 r^3 \sin 3\theta +$$

$$+ \Delta_1 r^4 \cos 4\theta + Q_1 r^4 \sin 4\theta \, .$$

For a long quadrupole (two-dimensional approximation), the ideal pole-pieces are given by $\varphi_1 = {}^1/_2\, Q r^2 \sin 2\theta$, in which φ_1 is the potential at the pole; Fig. 8 shows the pole-profiles when each of G_1, H_1, Q_1 and Δ_1 is superimposed upon the ideal shape. For each profile, *Deltrap* considers the shift of the point $\mathrm{d}r/\mathrm{d}\theta = 0$, for

$r = a$ and $\theta = \pi/4 + \delta\,\theta$; setting $Q = 2\,\varphi_1/a^2$ and $\delta\,\theta = \delta_3/a$ (δ_3 is tangential shift due to a term in 3θ), we find $G_1 = -4\sqrt{2}\,\delta_3\,\varphi_1/a^4$, $H_1 = 4\sqrt{2}\,\delta_3\,\varphi_1/a^4$. For the $\sin 4\theta\,(\delta_{4s})$ term, $Q_1 = -\delta_{4s}\,\varphi_1/a^5$; for the $\cos 4\theta$ term, we write $\theta = \pi/4$, $r = a +$

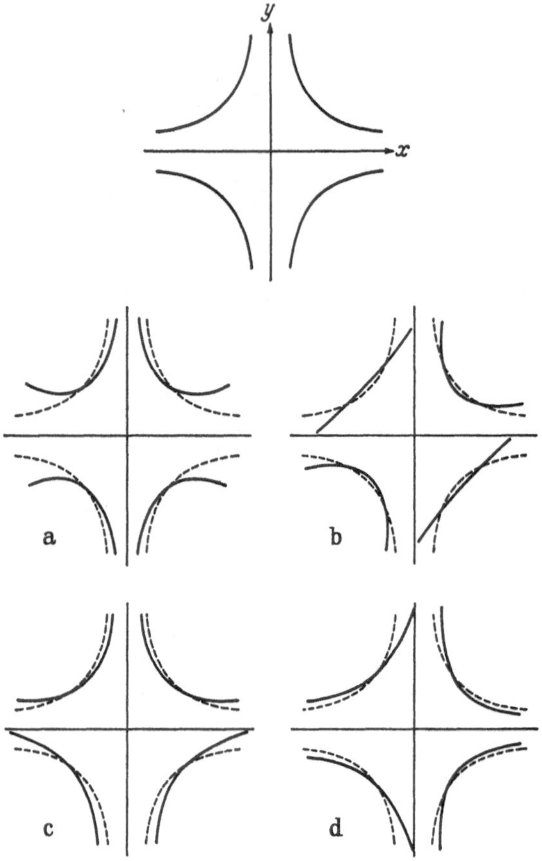

Fig. 8. The various types of pole-piece defect. (By courtesy of Dr. *Deltrap*)

$+\ \delta_{4c}$, giving $\Delta_1 = 2\,\delta_{4c}\,\varphi_1/a^5$. For short quadrupoles, φ_1 is replaced by $s\,\varphi_1$, and the functions $G_1 \ldots \Delta_1$ are written

$$G_1 = -4\sqrt{2}\,\frac{q\,\delta_3\,s\,\varphi_1}{a^4}\,k(z)\,; \quad H_1 = 4\sqrt{2}\,\frac{q\,\delta_3\,s\,\varphi_1}{a^4}\,k(z)$$

$$Q_1 = -\frac{q\,\delta_{4s}\,s\,\varphi_1}{a^5}\,k(z)\,; \quad \Delta_1 = 2\,\frac{q\,\delta_{4c}\,s\,\varphi_1}{a^5}\,k(z)$$

where

$$Q = \frac{2s\,\varphi_1}{a^2}\,k(z)$$

and q is a correction factor which is designed to make some allowance for differences between the z-dependence of Q and the parasitic functions.

The resulting aberrations are of the form

$$x^{(p)} = C_{20}\,\alpha^2 + C_{11}\,\alpha\,\beta + C_{02}\,\beta^2 + C_{30}\,\alpha^3 + C_{21}\,\alpha^2\,\beta + \\ + C_{12}\,\alpha\,\beta^2 + C_{03}\,\beta^3$$

in which α and β denote slopes in the current plane, $\alpha = x_a\,h'_{xe}$, $\beta = y_a\,h'_{ye}$. The coefficients are given by

$$C_{20} = 2\sqrt{2}\,\varkappa\,\delta_3 \int\limits_0^c h_x^3\,k\,\mathrm{d}z$$

$$C_{11} = 4\sqrt{2}\,\varkappa\,\delta_3 \int\limits_0^c h_x^2\,h_y\,k\,\mathrm{d}z$$

$$C_{02} = 2\sqrt{2}\,\varkappa\,\delta_3 \int\limits_0^c h_x\,h_y^2\,k\,\mathrm{d}z$$

$$C_{30} = 2\,\varkappa\,\delta_{4s} \int\limits_0^c h_x^4\,k\,\mathrm{d}z/a$$

$$C_{12} = 6\,\varkappa\,\delta_{4s} \int\limits_0^c h_x^2\,h_y^2\,k\,\mathrm{d}z/a$$

$$C_{21} = 12\,\varkappa\,\delta_{4c} \int\limits_0^c h_x^3\,h_y\,k\,\mathrm{d}z/a$$

$$C_{03} = 4\,\varkappa\,\delta_{4c} \int\limits_0^c h_x\,h_y^3\,k\,\mathrm{d}z/a$$

with $\varkappa = 2\eta\,q\,s\,\varphi_1/a^4\sqrt{\Phi}\,h'_c$. The coefficients C_{30} and C_{12} are measures of the familiar aperture aberrations associated with an octopole (cf. [129]); the remaining five represent parasitic aberrations.

Deltrap concludes with a comparison between the effects of the geometrical and parasitic aberrations. In the direction $x_a = y_a$, he finds

$$\frac{x^{(p)}}{x^{(3)}} = \frac{E_{20}}{E_2\,x_a} + \frac{E_{11}}{E_2\,x_a} + \frac{E_{02}}{E_2\,x_a} + \frac{E_{21}}{E_2} + \frac{E_{03}}{E_2}$$

in which $E_{ij} = C_{ij}/f_x^i\,f_y^j$ and $E_2 = C_2/f_x\,f_y^2$ (C_2 is the aperture aberration coefficient associated with $\alpha\,\beta^2$). He plots curves showing the dependence of $(E_{20} + E_{11} + E_{02})\,a^2/q\,\delta_3\,L^2\,E_2$ upon f_x/L and of $(E_{21} + E_{03})\,a^3/q\,\delta_4\,L^2$ upon f_x/L, and gives some idea of the practical values of the consequent tolerances, δ; for $f_x/L = 2$, $q = 0.3$ and $L = 3$ cm*, for example, he finds:

a (mm)	δ (mm)	$\frac{E_{21}+E_{03}}{E_2}$	$\frac{E_{20}+E_{11}+E_{02}}{E_2\,x_a}$		
			$\alpha = 0.1$	0.01	0.001
20	2	1.2	2.3	23	230
20	0.02	0.012	0.023	0.23	2.3
20	0.02	0.012			
15	0.02	0.03			
10	0.02	0.11			
5	0.02	0.88			

c) A. V. Crewe's study of the stigmatic magnetic doublet [36]. In this note, the effects of a small difference in azimuth between the poles of one quadrupole and those of the other, of a small tilt at one of the quadrupoles, and of a parallel displacement of the optic axes of the quadrupoles, are considered; *Crewe* demonstrates

* L denotes the equivalent length, given by the formula, $L = l + 1.1\,a$, l = mechanical length, a = bore radius.

that the last two defects merely produce linear shifts, and do not blur the image point, but that the first [which introduces a small component $\Delta(z)$] does produce an astigmatism.* It is suggested that the latter may perhaps be corrected by inserting a weak quadrupole between the components of the doublet, inclined to the correctly orientated member at 45° [and thus producing a pure $\Delta(z)$ component].

(The work of *Crewe* and *Meads* [131] is briefly summarized in *Cohen* [31].)

d) P. F. Meads' analysis of any failure to achieve quadrupole symmetry [130, 131]. Setting $\mathfrak{Q} = \eta\, Q/\sqrt{\Phi}$, $\nu(z) = -\eta\, G/\sqrt{\Phi}$, $\mu(z) = -\eta\, H/\sqrt{\Phi}$, $\Omega(z) = -\eta\, \Phi_m/\sqrt{\Phi}$, $\mathfrak{Q}_1(z) = 12\eta\, Q_1/\sqrt{\Phi}$, $\lambda(z) = \eta\, \Delta/4\sqrt{\Phi}$, *Glaser's* general formula for φ^m becomes

$$\begin{aligned}\frac{\eta}{\sqrt{\Phi}}\varphi^m = &-\Omega + \nu x + \mu y + \lambda(x^2 - y^2) + \\ &+ (\mathfrak{Q} + \delta\mathfrak{Q})\, x y - {}^1/_{12}\,\mathfrak{Q}''\, x y (x^2 + y^2) + \\ &+ {}^1/_3\,\mathfrak{Q}_1\, x y (x^2 - y^2)\end{aligned}$$

(*Meads* does not include the 3θ-terms, stating that these would not result from a simple displacement or rotation).

The equations of motion are thus

$$\begin{aligned}x'' + \mathfrak{Q} x = {}&{}^1/_{12}\,\mathfrak{Q}'' x^3 + {}^1/_4\,\mathfrak{Q}'' x y^2 + \mathfrak{Q}' x y y' - {}^3/_2\,\mathfrak{Q} x x'^2 - \\ &- {}^1/_2\,\mathfrak{Q} x y'^2 + \mathfrak{Q} y x' y' - {}^1/_3\,\mathfrak{Q}_1 x^3 + \mathfrak{Q}_1 x y^2 - x\,\delta\mathfrak{Q} + \\ &+ 2\lambda y - \mu \\ y'' - \mathfrak{Q} y = {}&- {}^1/_{12}\,\mathfrak{Q}'' y^3 - {}^1/_4\,\mathfrak{Q}'' x^2 y - \mathfrak{Q}' x y x' + {}^3/_2\,\mathfrak{Q} y y'^2 + \\ &+ {}^1/_2\,\mathfrak{Q} y x'^2 - \mathfrak{Q} x x' y - {}^1/_3\,\mathfrak{Q}_1 y^3 + \mathfrak{Q}_1 x^2 y + y\,\delta\mathfrak{Q} + \\ &+ 2\lambda x + \nu .\end{aligned}$$

The new terms in these equations lead to parasitic aberrations, $x^{(p)}$, $y^{(p)}$, given by

$$\begin{aligned}x^{(p)} &= -\int^z \delta\mathfrak{Q}\,\mathfrak{g}\, x\, d\zeta - \int^z \mathfrak{g}\,\mu\, d\zeta + 2\int^z \lambda\,\mathfrak{g}\, y\, d\zeta \\ y^{(p)} &= \int^z \delta\mathfrak{Q}\,\mathfrak{g}\, y\, d\zeta - \int^z \mathfrak{g}\,\nu\, d\zeta + 2\int^z \lambda\,\mathfrak{g}\, x\, d\zeta\end{aligned}$$

and hence, by

$$\begin{aligned}x^{(p)} &= \frac{\mu}{\mathfrak{Q}} T_1^0 + \frac{\delta\mathfrak{Q}}{\mathfrak{Q}} T_1^1\, x_o + \frac{\lambda}{\mathfrak{Q}} T_1^2\, y_o + \frac{\delta\mathfrak{Q}}{\mathfrak{Q}} T_1^3\, x_o' + \frac{\lambda}{\mathfrak{Q}} T_1^4\, y_o' \\ y^{(p)} &= \frac{\nu}{\mathfrak{Q}} T_2^0 + \frac{\lambda}{\mathfrak{Q}} T_2^1\, x_o + \frac{\delta\mathfrak{Q}}{\mathfrak{Q}} T_2^2\, y_o + \frac{\lambda}{\mathfrak{Q}} T_2^3\, x_o' + \frac{\delta\mathfrak{Q}}{\mathfrak{Q}} T_2^4\, x_o'\end{aligned}$$

in which the "tolerance coefficients" T_i^j on a quadrupole are given by the following integrals, taken over the corresponding quadrupole region:

$$\begin{aligned}&T_1^0 = s_x \int \mathfrak{Q}\, T_x\, dz - T_x \int \mathfrak{Q}\, s_x\, dz; \quad T_2^0 = -s_y \int \mathfrak{Q}\, T_y\, dz + T_y \int \mathfrak{Q}\, s_y\, dz \\ &T_1^1 = s_x \int \mathfrak{Q}\, s_x T_x\, dz - T_x \int \mathfrak{Q}\, s_x^2\, dz; \quad T_2^1 = 2 s_y \int \mathfrak{Q}\, s_x T_y\, dz - 2 T_y \int \mathfrak{Q}\, s_x s_y\, dz \\ &T_1^2 = -2 s_x \int \mathfrak{Q}\, s_y T_x\, dz + 2 T_x \int \mathfrak{Q}\, s_x s_y\, dz; \quad T_2^2 = -s_y \int \mathfrak{Q}\, s_y T_y\, dz + \\ &\qquad\qquad + T_y \int \mathfrak{Q}\, s_y^2\, dz \\ &T_1^3 = s_x \int \mathfrak{Q}\, T_x^2\, dz - T_x \int \mathfrak{Q}\, s_x T_x\, dz; \quad T_2^3 = 2 s_y \int \mathfrak{Q}\, T_x T_y\, dz - 2 T_y \int \mathfrak{Q}\, s_y T_x\, dz \\ &T_1^4 = -2 s_x \int \mathfrak{Q}\, T_x T_y\, dz + 2 T_x \int \mathfrak{Q}\, s_x T_y\, dz; \quad T_2^4 = -s_x \int \mathfrak{Q}\, T_y^2\, dz + \\ &\qquad\qquad + T_y \int \mathfrak{Q}\, s_y T_y\, dz .\end{aligned}$$

---

* The magnitude of this defect in an electrostatic quadrupole objective is discussed by *Bauer* [10]. *Crewe*'s conclusion is in agreement with a verbal comment made by *J. P. Davey* at the Cambridge Conference on Non-conventional Electron Microscopy (1965) and during earlier discussion. [103a] is concerned wholly with doublet defects.

e) Comment. It is clear that, despite the attempts by *Deltrap* to relate deformations to potential components, what is lacking in the theoretical work on tolerances is a reliable guide to the magnitudes and distributions of these parasitic components for a wide range of quadrupole shapes and associated defects. Furthermore, although it is usual to correct the lower order defects before attempting to cope with those of higher order, the relative importance of the various kinds of defects needs more careful study than it has as yet received.

4. Values of the cardinal elements and aberration coefficients of quadrupole lenses

4.1. Potential models

The functions $\Phi(z)$, $D(z)$ and $Q(z)$, in terms of which all the properties of twist-free systems are expressed, can often be replaced in practice by analytical expressions. The most convenient models are those which yield paraxial equations of motion possessing solutions in terms of tabulated functions, but if a computer is used to solve the equations, it is still very advantageous to use analytical expressions to avoid numerical differentiation. With long quadrupoles, the rectangular model is commonly employed for $D(z)$ and $Q(z)$, and for greater accuracy these rectangular potentials may be terminated by a fringe field. With the rectangular model, $D(z)$ or $Q(z)$ is constant over a region $a < z < b$, and vanishes outside these limits (Fig. 9a). This is the model commonly

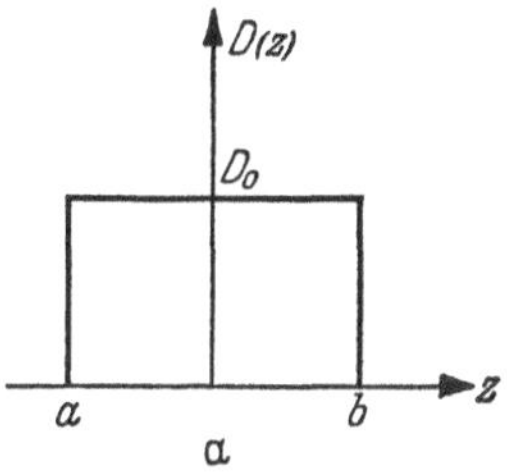

Fig. 9a. The rectangular model

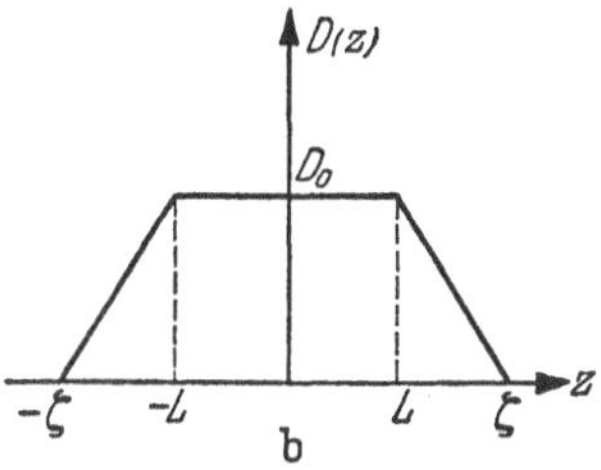

Fig. 9b. The triangular model

used in studies of strong-focusing lenses (see [182], or the reviews listed in § 1). Another fairly crude model is the triangular model [196, 92], in which $D(z)$ or $Q(z)$ consists of two straight sloping segments separated by a central plateau (Fig. 9b). For short quadrupoles, we return to the equations of motion (2.18), and substitute $z = d\cot\varphi$, $x = u\operatorname{cosec}\varphi$, $y = v\operatorname{cosec}\varphi$ and $\zeta = \sqrt{\Phi_0}\int_0^{\varphi}\Phi^{-1/2}\,d\varphi$, giving

$$\begin{aligned}
&\frac{d^2 u}{d\zeta^2} + \frac{1}{\Phi_0}\left\{\Phi + \frac{1}{4}\frac{d^2\Phi}{d\varphi^2} - \frac{d^2}{4\sin^4\varphi}\left(D - 4\eta\, Q\sqrt{\Phi}\right)\right\} u = 0 \\
&\frac{d^2 v}{d\zeta^2} + \frac{1}{\Phi_0}\left\{\Phi + \frac{1}{4}\frac{d^2\Phi}{d\varphi^2} + \frac{d^2}{4\sin^4\varphi}\left(D - 4\eta\, Q\sqrt{\Phi}\right)\right\} v = 0\,.
\end{aligned} \tag{4.1}$$

The solutions of these equations will be circular or hyperbolic functions if the coefficients of u and v are constants:

$$\begin{aligned}\frac{d^2\Phi}{d\varphi^2} + 4\Phi - d^2(D - 4\eta Q\sqrt{\Phi})\operatorname{cosec}^4\varphi = 4\Phi_o\,\omega_x^2\\ \frac{d^2\Phi}{d\varphi^2} + 4\Phi + d^2(D - 4\eta Q\sqrt{\Phi})\operatorname{cosec}^4\varphi = 4\Phi_o\,\omega_y^2\end{aligned} \tag{4.2}$$

so that

$$\frac{d^2\Phi}{d\varphi^2} + 4\Phi = 4\Phi_o\,\mu^2\,,\quad 2\mu^2 = \omega_x^2 + \omega_y^2 \tag{4.3a}$$

$$D - 4\eta Q\sqrt{\Phi} = 4\Phi_o\,k^2\sin^4\varphi/d^2\,,\quad 2k^2 = \omega_y^2 - \omega_x^2\,. \tag{4.3b}$$

From equation (4.3a), we find

$$\begin{aligned}\Phi(\varphi) = \Phi_o(1 - \varkappa^2\sin^2\varphi)\qquad \varkappa^2 = -\frac{\Phi(\pi/2) - \Phi_o}{\Phi_o}\\ \mu^2 = 1 - {}^1\!/_2\,\varkappa^2\end{aligned} \tag{4.4}$$

which is the bell-shaped field of *Glaser* and *Schiske* [72, 160], eminently suitable for round electrostatic einzel lenses.

Equation (4.3b), with $k^2 = k_D^2 + k_Q^2$, yields

$$D(\varphi) = \frac{4\Phi_o\,k_D^2}{d^2}\sin^4\varphi\,,\quad Q(\varphi) = -\frac{\Phi_o\,k_Q^2}{\eta\, d^2\sqrt{\Phi}}\sin^4\varphi \tag{4.5a}$$

or

$$\begin{aligned}D(z) = \frac{D_0}{\{1 + (z/d)^2\}^2}\quad \text{with}\quad D_0 = 4\Phi_o\,k_D^2/d^2\\ Q(z) = \frac{Q_0}{\{1 + (z/d)^2\}^2}\quad \text{with}\quad \eta Q_0 = -k_Q^2\sqrt{\Phi_o}\,/d^2\end{aligned} \tag{4.5b}$$

provided that $\Phi = \Phi_o$. If this last condition is not satisfied,

$$Q(\varphi) = -\frac{k_Q^2\sqrt{\Phi_o}}{\eta\, d^2}\,\frac{\sin^4\varphi}{1 - \varkappa^2\sin^2\varphi}\,. \tag{4.6}$$

All these models provide twist-free systems with $\Theta = 0$. *Dušek* [45] has shown that a torus with an even number of breaks offers a means of obtaining twist-free systems for which $\Theta \neq 0$. His analytic solution of the corresponding $2n$-electrode lens is valid only when the breaks are narrow enough for a linear approximation to the potential across them to be legitimate. In the electrostatic case, we have $\tan 2\Theta = 2P/D$ from the orthogonality condition, and if Θ is constant, we write $\tan 2\Theta = K$, $D = 2P/K$. For a $2n$-electrode torus, with potential U_m on the m-th electrode and equal potentials on opposite electrodes, *Dušek* finds

$$\begin{aligned}K = \Bigg[\sum_1^n U_m\left\{\sin\frac{\pi}{n}(2m-1)\sin\left(\frac{\pi}{n} - 2\delta\right) + \sin\frac{2\pi}{n}m\sin 2\delta\right\} +\\ + \frac{U_{m+1} - U_m}{2}\left\{\frac{\sin 2\delta}{2\delta}\cos 2\pi\frac{m}{n} - \cos\left(\frac{2\pi}{n}m + 2\delta\right)\right\}\Bigg]\Bigg/\\ \Bigg[\sum_1^n U_m\left\{\cos\frac{\pi}{n}(2m-1)\sin\left(\frac{\pi}{n} - 2\delta\right) + \cos\frac{2\pi}{n}m\sin 2\delta\right\} -\\ - \frac{U_{m+1} - U_m}{2}\left\{\frac{\sin 2\delta}{2\delta}\sin\frac{2\pi}{n}m - \sin\left(\frac{2\pi}{n}m + 2\delta\right)\right\}\Bigg]\end{aligned}$$

(see Fig. 10) and for $\delta \simeq 0$, this simplifies to

$$K = \frac{\sum_1^n U_m \sin \frac{\pi}{n}(2m-1)}{\sum_1^n U_m \cos \frac{\pi}{n}(2m-1)}.$$

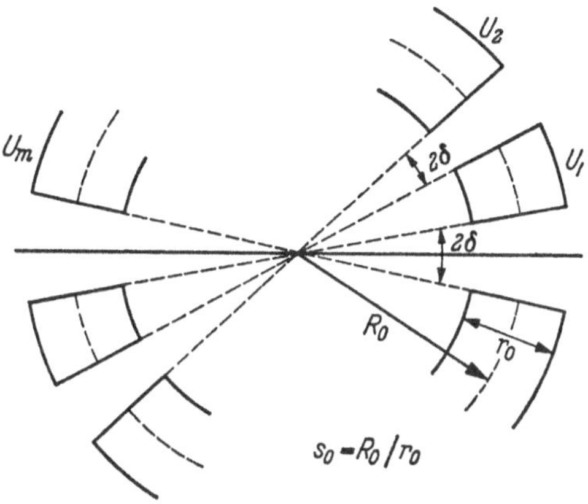

Fig. 10. The torus as $2n$-pole. (After *Dušek* [45])

Thus for $n = 2$, we obtain a quadrupole,

$$K = \tan 2\Theta \to \infty, \quad \Theta = \pi/4, \quad D \equiv 0.$$

That is, the original potential is of the form

$$\varphi(X, Y, z) = \Phi - {}^1\!/_4\, \Phi''(X^2 + Y^2) + PXY + \cdots$$

and these axes must be rotated through $\pi/4$ to obtain a coordinate

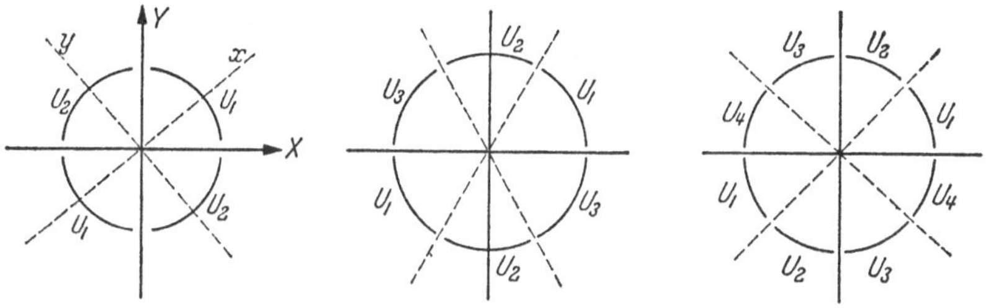

Fig. 11. The torus as quadrupole, sextupole or octopole. (After *Dušek* [45])

system (x, y, z) in which the equations of motion separate:

$$\varphi(x, y, z) = \Phi - {}^1\!/_4\, \Phi''(x^2 + y^2) + {}^1\!/_2\, P(x^2 - y^2)$$

(see Fig. 11).

For $n = 3$, $$K = \tan 2\Theta = \frac{U_1 - U_3}{U_1 - 2U_2 + U_3}\sqrt{3};$$

with $U_1 = -U_3$, $\tan 2\alpha = -\sqrt{3}\, U_1/U_2$ and with $U_1 + U_2 + U_3 = 0$, $\tan 2\alpha = -(U_1 - U_3)/\sqrt{3}(U_1 + U_3)$: such a sextupole behaves like an inclined quadrupole.

For $n = 4$,

$$K = \tan 2\Theta \frac{(U_1 + U_2) - (U_3 + U_4)}{(U_1 + U_4) - (U_2 + U_3)}$$

and if $U_1 = U_4$ and $U_2 = U_3$, $K = \Theta = 0$ and the octopole behaves like a symmetrically orientated quadrupole together with a round lens.

4.2. Transfer matrices and cardinal elements

a) The rectangular model. With a symmetrical quadrupole, the equations of motion are of the form

$$x'' - \beta^2 x = 0 \quad y'' + \beta^2 y = 0 \tag{4.7}$$

in which

$$\beta^2 = \frac{D_0 - 4\eta\, Q_0 \sqrt{\Phi}}{4\Phi}\,. \tag{4.8}$$

We can thus solve the equations of motion through isolated electrostatic and magnetic quadrupoles, and also, through overlapping electrostatic and magnetic quadrupoles. Consider first an isolated quadrupole. $\beta^2 = 0$ for $|z| > \zeta$. Within the lens,

$$x = \left\{x(-\zeta)\,\mathrm{Sh}\,\beta\zeta + \frac{x'(-\zeta)}{\beta}\,\mathrm{Ch}\,\beta\zeta\right\}\mathrm{Sh}\,\beta z +$$

$$+ \left\{x(-\zeta)\,\mathrm{Ch}\,\beta\zeta + \frac{x'(-\zeta)}{\beta}\,\mathrm{Sh}\,\beta\zeta\right\}\mathrm{Ch}\,\beta z$$

$$y = \left\{-y(-\zeta)\sin\beta\zeta + \frac{y'(-\zeta)}{\beta}\cos\beta\zeta\right\}\sin\beta z +$$

$$+ \left\{y(-\zeta)\cos\beta\zeta + \frac{y'(-\zeta)}{\beta}\sin\beta\zeta\right\}\cos\beta z$$

and the transfer matrices are thus

$$\begin{pmatrix} x(\zeta) \\ p(\zeta) \end{pmatrix} = \begin{pmatrix} \mathrm{Ch}\,2\beta\zeta & \dfrac{\mathrm{Sh}\,2\beta\zeta}{\beta\sqrt{\Phi}} \\ \sqrt{\Phi}\,\beta\,\mathrm{Sh}\,2\beta\zeta & \mathrm{Ch}\,2\beta\zeta \end{pmatrix} \begin{pmatrix} x(-\zeta) \\ p(-\zeta) \end{pmatrix} \tag{4.9a}$$

$$\begin{pmatrix} y(\zeta) \\ q(\zeta) \end{pmatrix} = \begin{pmatrix} \mathrm{Cos}\,2\beta\zeta & \dfrac{\sin 2\beta\zeta}{\beta\sqrt{\Phi}} \\ -\beta\sqrt{\Phi}\sin 2\beta\zeta & \cos 2\beta\zeta \end{pmatrix} \begin{pmatrix} y(-\zeta) \\ q(-\zeta) \end{pmatrix}. \tag{4.9b}$$

The image focal lengths, foci and principal points are thus [cf. Equation (2.13)]

$$\begin{gathered} f_{xi} = 1/\beta\,\mathrm{Sh}\,2\beta\zeta \quad f_{yi} = -1/\beta\sin 2\beta\zeta \\ z_{Fi}^{(x)} = \zeta - \frac{\coth 2\beta\zeta}{\beta} \quad z_{Fi}^{(y)} = \zeta + \frac{\cot 2\beta\zeta}{\beta} \\ z_{Hi}^{(x)} = \zeta - \frac{\mathrm{th}\,\beta\zeta}{\beta} \quad z_{Hi}^{(y)} = \zeta - \frac{\tan\beta\zeta}{\beta}\,. \end{gathered} \tag{4.10}$$

In the plane $z = 0$, we have

$$\varphi(x, y, 0) = \Phi_0 + {}^1\!/_4\, D_0 (x^2 - y^2) - \cdots$$

for an electrostatic quadrupole, so that to a first approximation,

$$D_0 = \frac{4\Phi_Q}{a^2}, \quad \beta^2 = \frac{\Phi_Q}{a^2 \Phi_0} \tag{4.11}$$

in which Φ_Q denotes the applied potential and the electrodes are distance a from the axis. The argument $2\beta\zeta$ can thus be written

$$2\beta\zeta = \frac{2\zeta}{a}\sqrt{\frac{\Phi_Q}{\Phi_0}}\,.$$

The model is only a good one when $2\zeta \gg a$, and hence the weak lens approximation ($\sin 2\beta\zeta = \mathrm{Sh}\,2\beta\zeta \simeq 2\beta\zeta$; $\cos 2\beta\zeta = \mathrm{Ch}\,2\beta\zeta \simeq 1$) is unlikely to describe many practical situations very accurately.

The excitation is occasionally applied to the electrodes asymmetrically [41–43, 180–184], so that the applied potentials are $\Phi_x(x = \pm a, y = 0)$ and $-\Phi_y(y = \pm b, x = 0)$; β^2 now becomes

$$\beta^2 = \frac{\Phi_x + \Phi_y}{\left(\Phi_0 + \dfrac{\Phi_x - \Phi_y}{2}\right)(a^2 + b^2)}$$

in the electrostatic case, which reduces to *Dhuicq's* expressions for $\Phi_x = 0$, $\Phi_y = 2\Phi_i$, $a = b$. The octopole component thus introduced may have a beneficial effect upon the aberrations [184, 53, 212].

b) The bell-shaped model

(i) Φ constant.

The potential is now of the form

$$D(z) = \frac{D_0}{\{1 + (z/d)^2\}^2} \quad Q(z) = \frac{Q_0}{\{1 + (z/d)^2\}^2} \tag{4.12}$$

and the equations of motion can only be solved in superimposed electrostatic and magnetic quadrupoles if the latter are symmetrical about the same plane $z = 0$ and fall off at the same rate (so that the parameter d is the same for both). We consider the electrostatic case, but D may be replaced by $D - 4\eta Q\sqrt{\Phi}$ to obtain the magnetic or mixed cases. The equations of motion are now

$$\frac{d^2 u}{d\varphi^2} + (1 - k^2)\,u = 0 \quad \frac{d^2 v}{d\varphi^2} + (1 + k^2)\,v = 0 \tag{4.13}$$

with $k^2 = D_0 d^2/4\Phi$ [cf. Equations (4.5)], and asymptotically,

$$\begin{pmatrix} x_c \\ p_c \end{pmatrix} = \begin{pmatrix} \dfrac{z_c \sin\pi\omega_x - d\omega_x\cos\pi\omega_x}{d\omega_x} & \dfrac{(z_0 - z_c)\cos\pi\omega_x - (d^2\omega_x^2 + z_0 z_c)\sin\pi\omega_x/d\omega_x}{\sqrt{\Phi}} \\ \sqrt{\Phi}\,\dfrac{\sin\pi\omega_x}{d\omega_x} & -\dfrac{z_0\sin\pi\omega_x + d\omega_x\cos\pi\omega_x}{d\omega_x} \end{pmatrix} \begin{pmatrix} x_0 \\ p_0 \end{pmatrix} \tag{4.14}$$

in which $\omega_x^2 = 1 - k^2$; the y-transfer matrix is obtained by writing ω_y for ω_x, $\omega_y^2 = 1 + k^2$. The cardinal elements are thus given by

$$f_{xi} = \frac{d\omega_x}{\sin\pi\omega_x} \quad z_{Fi}^{(x)} = d\omega_x\cot\pi\omega_x$$
$$z_{Hi}^{(x)} = d\omega_x\cot\frac{\pi\omega_x}{2} \tag{4.15a}$$

$$f_{yi} = \frac{d\,\omega_y}{\sin\pi\,\omega_y} \qquad z_{Fi}^{(y)} = d\,\omega_y \cot\pi\,\omega_y$$
$$z_{Hi}^{(y)} = d\,\omega_y \cot\frac{\pi\,\omega_y}{2}\,. \tag{4.15b}$$

If $k^2 > 1$, then $\omega_x^2 < 0$ and the x-cardinal elements become

$$f_{xi} = \frac{d\sigma_x}{\mathrm{Sh}\,\pi\,\sigma_x} \qquad z_{Fi}^{(x)} = d\sigma_x \coth\pi\,\sigma_x$$
$$z_{Hi}^{(x)} = d\sigma_x \coth\frac{\pi\,\sigma_x}{2} \tag{4.15c}$$

in which $\sigma_x^2 = k^2 - 1$.

Tanguy [218] gives formulae for the cardinal elements as a function of object distance (the immersion cardinal elements). For $z \geqq 3d$, these are scarcely distinguishable from the asymptotic values.

(ii) $\Phi = \Phi(z)$.

The equations of motion are still soluble in terms of tabulated functions if $\Phi(z)$, $D(z)$ and $Q(z)$ are all different from zero, provided that they can all be represented by bell-shaped distributions to a good approximation. Recapitulating, we now have

$$\frac{\mathrm{d}^2 u}{\mathrm{d}\zeta^2} + \omega_x^2\, u = 0 \qquad \frac{\mathrm{d}^2 v}{\mathrm{d}\zeta^2} + \omega_y^2\, v = 0$$

with $\zeta = \sqrt{\Phi_o}\int\limits_0^{\varphi}\frac{\mathrm{d}\varphi}{\sqrt{\Phi}}$, $\Phi(\varphi) = \Phi_o(1 - \varkappa^2\sin^2\varphi)$ and $\omega_x^2 = 1 - k^2 - {}^1/_2\,\varkappa^2$, $\omega_y^2 = 1 + k^2 - {}^1/_2\,\varkappa^2$; as before, $D - 4\eta\, Q\sqrt{\Phi} = (4\Phi_o/d^2)\, k^2 \sin^4\varphi$ $(k^2 = k_D^2 + k_Q^2)$. Denoting ζ by $F(\varphi, \varkappa)$, the following functions will be lienarly independent solutions of the equations of motion:

$$x(\varphi) = \frac{\sin\omega_x F(\varphi,\varkappa)}{\sin\varphi}\,; \quad x(\varphi) = \frac{\cos\omega_x F(\varphi,\varkappa)}{\sin\varphi}\,;$$
$$y(\varphi) = \frac{\sin\omega_y F(\varphi,\varkappa)}{\sin\varphi}\,; \quad y(\varphi) = \frac{\sin\omega_y F(\varphi,\varkappa)}{\sin\varphi}\,.$$

The imagery of such an element has been studied in detail by *Dušek* [45, 46], a part of whose analysis is summarized below. It possesses the interesting property that a single element can produce stigmatic and orthomorphic imagery, as we shall see. We note that if $\omega_x = 0$,

$$x(\varphi) = \operatorname{cosec}\varphi\,; \quad x(\varphi) = F(\varphi,\varkappa)\operatorname{cosec}\varphi\,;$$

and that if $\omega_x^2 = -\sigma_x^2$,

$$x(\varphi) = \frac{\mathrm{Sh}\,\sigma_x F(\varphi,\varkappa)}{\sin\varphi}\,; \quad x(\varphi) = \frac{\mathrm{Ch}\,\sigma_x F(\varphi,\varkappa)}{\sin\varphi}\,;$$

we have $\omega_y^2 \geqq {}^1/_2$.

If ω_x^2 is positive, the planes $\varphi = \varphi_o$ and $\varphi = \varphi_{xi}$ will be conjugate for the $x - z$ plane if

$$F(\varphi_o, \varkappa) - F(\varphi_{xi}, \varkappa) = \frac{\pi}{\omega_x}\, n_x$$

and for the $y - z$ plane, φ_o and φ_{yi} will be conjugate if

$$F(\varphi_o, \varkappa) - F(\varphi_{yi}, \varkappa) = \frac{\pi}{\omega_y} n_y .$$

If the object point tends to infinity ($\varphi_o \to \pi$),

$$F(\varphi_{xi}, \varkappa) = 2K - \frac{\pi}{\omega_x} n_x \qquad K = {}^1/_2 F(\pi, \varkappa)$$

$$F(\varphi_{yi}, \varkappa) = 2K - \frac{\pi}{\omega_y} n_y .$$

If there is (at least) one real image,

$$\varphi_{xi} \geqq 0 \text{ and } \varphi_{yi} \geqq 0 , \text{ or } n_x \pi/\omega_x \leqq 2K , \ n_y \pi/\omega_y \leqq 2K .$$

Since $\omega_y^2 \geqq {}^1/_2$, there is always a real image in the $y - z$ plane; in the $x - z$ plane, however, there will be no real image if $\pi/\omega_x > 2K$, so that the boundary between real and virtual imagery is given by

$$(\pi/2K)^2 = 1 - k^2 - {}^1/_2 \varkappa^2 .$$

If ω_x^2 is zero or negative, the imagery in the $x - z$ plane will always be virtual.

In Fig. 12a, the boundary between convergent and divergent action is shown, and in Fig. 12b, "telescopic" rays in the $x - z$ plane, when ω_x has the limiting value, $\pi/2K$.

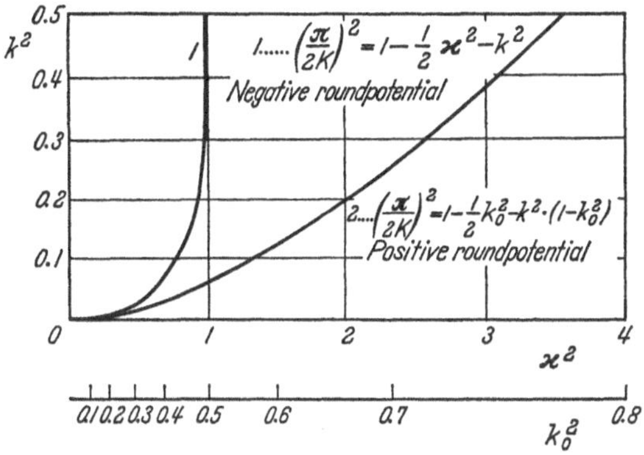

Fig. 12a. The boundary between convergent and divergent action in the $x - z$ plane

If the element is convergent in both directions, we may ask whether it can produce a stigmatic image, and whether the latter can also be orthomorphic. The magnifications M_x, M_y, satisfy the relations (cf. § 2)

$$M_x = \frac{\cos\omega_x F(\varphi_{xi}, \varkappa) \sin\varphi_{xo}}{\cos\omega_x F(\varphi_{xo}, \varkappa) \sin\varphi_{xi}} = \frac{\sin\omega_x F(\varphi_{xi}, \varkappa) \sin\varphi_{xo}}{\sin\omega_x F(\varphi_{xo}, \varkappa) \sin\varphi_{xi}}$$

$$M_y = \frac{\cos\omega_y F(\varphi_{yi}, k) \sin\varphi_{yo}}{\cos\omega_y F(\varphi_{yo}, \varkappa) \sin\varphi_{yi}} = \frac{\sin\omega_y F(\varphi_{yi}, \varkappa) \sin\varphi_{yo}}{\sin\omega_y F(\varphi_{yo}, \varkappa) \sin\varphi_{yi}}$$

and hence

$$M_x = (-1)^{n_x} \frac{\sin \varphi_{xo}}{\sin \varphi_{xi}}; \quad M_y = (-1)^{n_y} \frac{\sin \varphi_{yo}}{\sin \varphi_{yi}}$$

so that if $\varphi_{xi} = \varphi_{yi}$ when $\varphi_{xo} = \varphi_{yo}$, the magnifications are equal in magnitude: a stigmatic image is automatically an orthomorphic image.

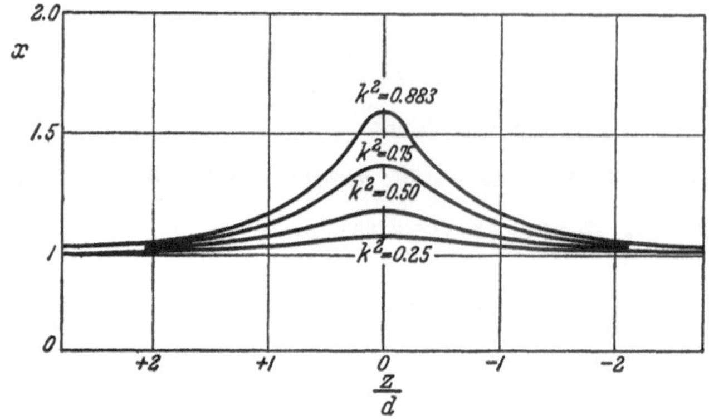

Fig. 12b. Telescopic rays in the $x - z$ plane

We consider only $n_x = 1$ and set the object plane at infinity, so that

$$\omega_y = n_y \, \omega_x$$

and

$$k^2 = \frac{n_y^2 - 1}{n_y^2 + 1} (1 - {}^1\!/_2 \, \varkappa^2) \, .$$

If $n_y = 1$, then $k = 0$ and the element is an ordinary round lens. For convergence in the $x - z$ plane,

$$\left(\frac{\pi}{2K}\right)^2 \leq \frac{2}{n_y^2 + 1} (1 - {}^1\!/_2 \, \varkappa^2)$$

and with $n_y = 2$, $\varkappa$ must lie between 0.9844 and 1. Thus 0.9844 is the smallest value of $\varkappa$ for which the equality sign holds, and the imagery is telescopic (Fig. 12c and d). If the object is a finite distance from the element, $\pi > \varphi_{xo} = \varphi_{yo} = \varphi_o > \frac{\pi}{2}$, we have only to replace $2K$ by $2K - F(\pi - \varphi_o, \varkappa)$ to obtain the new lower limit for $\varkappa$ (Fig. 12e). The trajectories through two elements, in which stigmatic imagery is and is not possible, are illustrated in Figs. 13a and b and 13c and d; in the former, $\varkappa$ is smaller than the minimum value deduced above: $\varkappa^2 = 0.587$; in the latter, $\varkappa$ is greater than the minimum: $\varkappa^2 = 0.992$ and stigmatic rays are shown ($k^2 = 0.302$).

When the round lens component is positive with respect to the accelerating voltage, we have

$$\zeta = \int_0^{\varphi} \frac{d\varphi}{\sqrt{1 + \varkappa^2 \sin^2 \varphi}} = F(\varphi, i\,\varkappa)$$

so that setting $k_0 = \varkappa/\sqrt{1+\varkappa^2}$, and $\Omega_1^2 = 1 - {}^1/_2\, k_0^2 - k^2(1-k_0^2)$, $\Omega_2^2 = 1 - {}^1/_2\, k_0^2 + k^2(1-k_0^2)$, the pairs of trajectories become

$$x(\varphi) = \frac{\cos\Omega_1(K+F)}{\sin\varphi}\,; \qquad x(\varphi) = \frac{\sin\Omega_1(K+F)}{\sin\varphi}\,;$$
$$y(\varphi) = \frac{\cos\Omega_2(K+F)}{\sin\varphi}\,; \qquad y(\varphi) = \frac{\sin\Omega_2(K+F)}{\sin\varphi}\,,$$

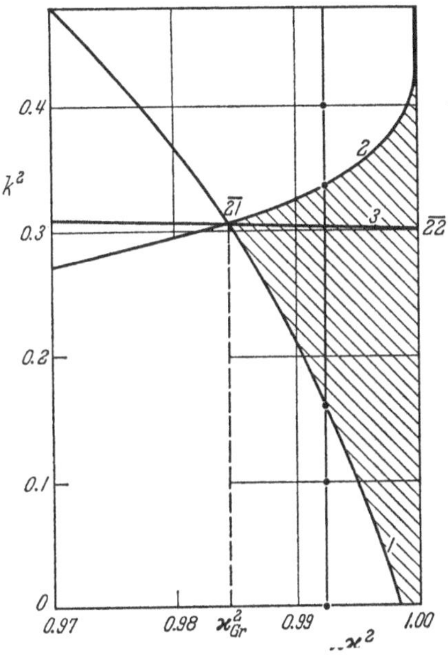

Fig. 12c

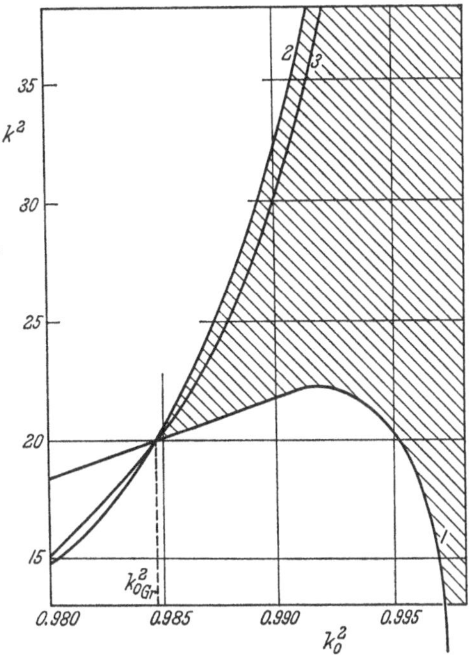

Fig. 12d

Fig. 12c and d. The pairs of values of k^2 and $\varkappa^2$ or k_0^2 for which the imagery is stigmatic lie on the line $\overline{21}$—$\overline{22}$

Fig. 12c. Negative round lens potential. Curve 1: $k^2 = (\pi/K)^2 + {}^1/_2\,\varkappa^2 - 1$, the boundary curve for the $y-z$ plane; double imagery occurs above the curve. Curve 2: $k^2 = 1 - {}^1/_2\,\varkappa^2 - (\pi/2K)^2$, the boundary curve for the $x-z$ plane; converging action corresponds to the region below the curve — in the barred zone, the $x-z$ plane is converging and the imagery in the $y-z$ plane is double. Curve 3: $k^2 = {}^3/_5\,(1 - {}^1/_2\,\varkappa^2)$, the values of $\varkappa^2 - k^2$ for which stigmatic imagery is possible

Fig. 12d. Positive round lens potential. Curve 1: $k^2 = \frac{1}{1-k_0^2}\left\{\left(\frac{\pi}{K}\right)^2 + \frac{k_0^2}{2} - 1\right\}$.

Curve 2: $k^2 = \frac{1}{1-k_0^2}\left\{1 - \frac{k_0^2}{2} - \left(\frac{\pi}{K}\right)^2\right\}$, convergent action to the right of the curve.

Curve 3: $k^2 = \frac{3}{5}\,\frac{1-k_0^2/2}{1-k_0^2}$

where K is the elliptic integral of the first kind with argument k_0 and F denotes $F(\varphi - \pi/2, k_0)$. The curves corresponding to this situation are also plotted in Figs. 12 and 13.

(iii) The unsymmetrical bell-shaped distribution.

The bell-shaped fields hitherto considered have all possessed symmetry about the centre plane. If this symmetry is abandoned, we write

$$D(z) = \frac{D_0}{\{1+(z/a)^2\}^2} \text{ for } z<0 \text{ and } D(z) = \frac{D_0}{\{1+(z/b)^2\}^2} \text{ for } z>0\,.$$

The transfer matrix in the $x - z$ plane is now given by the matrix product

$$\begin{pmatrix} \frac{z_o}{b}\cos\frac{\pi}{2}\theta_x + \theta_x \sin\frac{\pi}{2}\theta_x & \frac{z_o}{\sqrt{\Phi}\,\theta_x}\sin\frac{\pi}{2}\theta_x - \frac{b}{\sqrt{\Phi}}\cos\frac{\pi}{2}\theta_x \\ \frac{\sqrt{\Phi}}{b}\cos\frac{\pi}{2}\theta_x & \frac{1}{\theta_x}\sin\frac{\pi}{2}\theta_x \end{pmatrix} \times$$

$$\times \begin{pmatrix} \frac{1}{\omega_x}\sin\frac{\pi}{2}\omega_x & -\frac{z_o}{\sqrt{\Phi}\,\omega_x}\sin\frac{\pi}{2}\omega_x - \frac{a}{\sqrt{\Phi}}\cos\frac{\pi}{2}\omega_x \\ \frac{\sqrt{\Phi}}{a}\cos\frac{\pi}{2}\omega_x & -\frac{z_o}{a}\cos\frac{\pi}{2}\omega_x + \omega_x\sin\frac{\pi}{2}\omega_x \end{pmatrix}$$

with a similar expression in the $y - x$ plane.

$$\theta_x^2 = 1 - b^2\beta^2\,, \quad \omega_x^2 = 1 - a^2\beta^2\,,$$

$$\theta_y^2 = 1 + b^2\beta^2\,, \quad \omega_y^2 = 1 + a^2\beta^2\,.$$

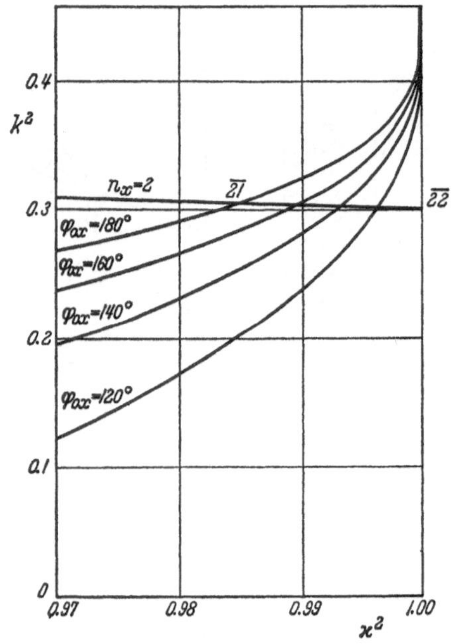

Fig. 12 e. Stigmatic and orthomorphic imagery for finite object distance. (After *Dušek* [45] and by courtesy of *Optik* [46])

The asymptotic cardinal elements are given by the following expressions

$$\frac{1}{f_{xi}} = -\frac{1}{f_{xo}} = \frac{\cos\frac{\pi}{2}\theta_x \sin\frac{\pi}{2}\omega_x}{b\,\omega_x} + \frac{\sin\frac{\pi}{2}\theta_x\cos\frac{\pi}{2}\omega_x}{a\,\theta_x}$$

$$z_{Fi} = f_{xi}\left(\frac{b}{a}\cos\frac{\pi}{2}\theta_x\cos\frac{\pi}{2}\omega_x - \frac{\theta_x}{\omega_x}\sin\frac{\pi}{2}\theta_x\sin\frac{\pi}{2}\omega_x\right)$$

$$z_{Fo} = f_{xo}\left(\frac{a}{b}\cos\frac{\pi}{2}\theta_x\cos\frac{\pi}{2}\omega_x - \frac{\omega_x}{\theta_x}\sin\frac{\pi}{2}\theta_x\sin\frac{\pi}{2}\omega_x\right)$$

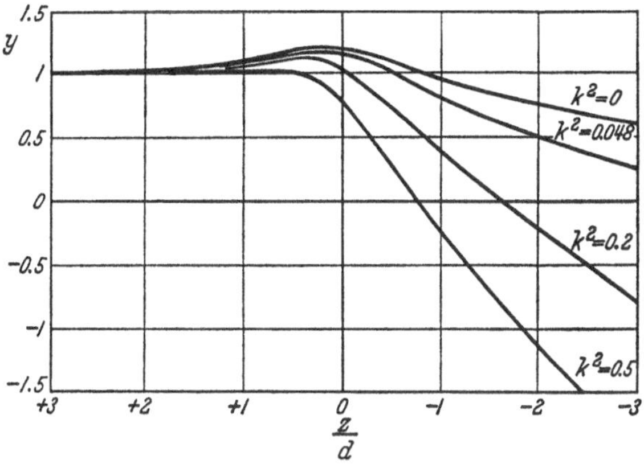

Fig. 13a

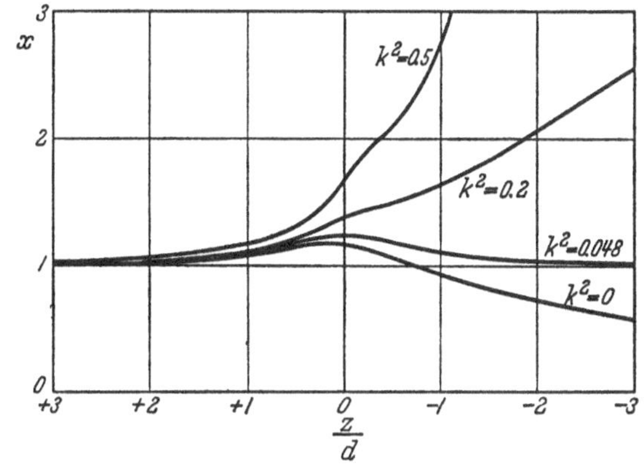

Fig. 13b

Fig. 13a and b. Rays in an element with $\varkappa^2 = 0.587$

which may be re-arranged thus:

$$\frac{1}{f_{xi}} = -\frac{1}{f_{xo}}$$

$$= \frac{1}{2}\left(\frac{1}{b\,\omega_x} + \frac{1}{a\,\theta_x}\right)\sin\frac{\pi}{2}(\omega_x + \theta_x) + \frac{1}{2}\left(\frac{1}{b\,\omega_x} - \frac{1}{a\,\theta_x}\right)\sin\frac{\pi}{2}(\omega_x - \theta_x)$$

$$z_{Fi} = \quad b\,\theta_x \frac{(a\,\theta_x + b\,\omega_x)\cos\frac{\pi}{2}(\omega_x + \theta_x) - (a\,\theta_x - b\,\omega_x)\cos\frac{\pi}{2}(\omega_x - \theta_x)}{(a\,\theta_x + b\,\omega_x)\sin\frac{\pi}{2}(\omega_x + \theta_x) + (a\,\theta_x - b\,\omega_x)\sin\frac{\pi}{2}(\omega_x - \theta_x)}$$

$$z_{Fo} = -a\,\omega_x \frac{(a\,\theta_x + b\,\omega_x)\cos\frac{\pi}{2}(\omega_x + \theta_x) - (a\,\theta_x - b\,\omega_x)\cos\frac{\pi}{2}(\omega_x - \theta_x)}{(a\,\theta_x + b\,\omega_x)\sin\frac{\pi}{2}(\omega_x + \theta_x) + (a\,\theta_x - b\,\omega_x)\sin\frac{\pi}{2}(\omega_x - \theta_x)}$$

We also list the "real" cardinal elements, as opposed to the asymptotic cardinal elements, for this case, from which the corresponding quantities for the symmetrical bell ($a = b$) are easily deduced. The asymptotic cardinal elements have been obtained by calculating the emergent asymptotes to the rays which are incident on the system with incoming

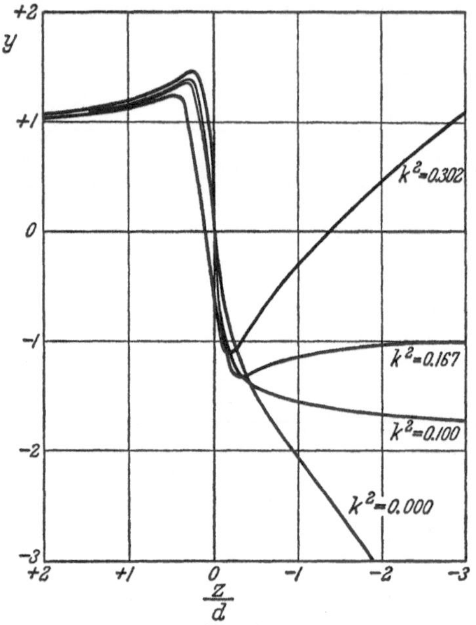

Fig. 13c

Fig. 13c and d. Rays in an element with $\varkappa^2 = 0.992$. (By courtesy of *Optik* [46])

asymptotes parallel to the optic axis. The "real" cardinal elements are given by the points at which these same rays intersect the axis, and their gradients at these points.

For the real cardinal elements, we find

$$\varphi_{Fo} = \frac{\pi}{2} - \frac{\arctan\left(\frac{b\,\omega_x}{a\,\theta_x}\tan\frac{\pi}{2}\,\theta_x\right) - n\,\pi}{\omega_x}$$

$$\psi_{Fi} = \frac{\pi}{2} + \frac{\arctan\left(\frac{a\,\theta_x}{b\,\omega_x}\tan\frac{\pi}{2}\,\omega_x\right) - n\,\pi}{\theta_x}$$

$$f_{xi} = \frac{a\,b\,\omega_x}{\sqrt{\left(a^2\,\theta_x^2\sin^2\frac{\pi}{2}\,\omega_x + b^2\,\omega_x^2\cos^2\frac{\pi}{2}\,\omega_x\right)}}\operatorname{cosec}\psi_{Fi}$$

$$f_{xo} = -\frac{a\,b\,\theta_x}{\sqrt{\left(a^2\,\theta_x^2\cos^2\frac{\pi}{2}\,\theta_x + b^2\,\omega_x^2\sin^2\frac{\pi}{2}\,\omega_x\right)}}\operatorname{cosec}\varphi_{Fo}$$

$$z_{Fi} = b\cot\psi_{Fi} \qquad z_{Fo} = a\cot\varphi_{Fo}\,.$$

For the symmetrical bell, therefore, for which $a = b = d$,

$$\varphi_{Fo} = \frac{n\pi}{\omega_x} \quad \psi_{Fi} = \varphi_{Fi} = \pi - \frac{n\pi}{\omega_x}$$

$$f_{xi} = d \operatorname{cosec} \frac{n\pi}{\omega_x} = -f_{xo}$$

$$z_{Fi} = -d \cot \frac{n\pi}{\omega_x} = -z_{Fo}\,.$$

c) The modified rectangular model. This is a model commonly used for a more exact analysis of strong focusing lenses than the rectangular model. The latter is now flanked by two fringe potentials, which may

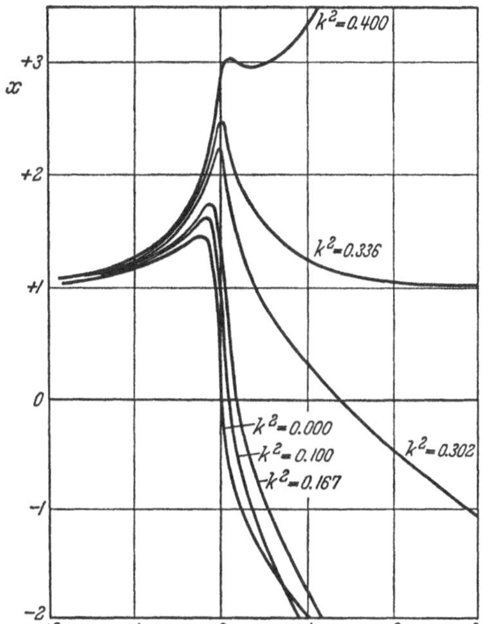

Fig. 13d

conveniently be assumed to be bell-shaped:

$$|z| < \zeta\,, \quad D = D_0\,;$$

$$z < -\zeta\,, \quad D(z) = \frac{D_0}{\{1 + [(z+\zeta)/d]^2\}^2}\,; \quad z > \zeta\,, \quad D(z) = \frac{D_0}{\{1 + [(z-\zeta)/d]^2\}^2}$$

The asymptotic transfer matrix in the $x - z$ plane is given by the matrix product

$$\begin{pmatrix} \frac{Z_c}{d}\cos\frac{\pi}{2}\omega_x + \omega_x\sin\frac{\pi}{2}\omega_x & \frac{Z_c}{\sqrt{\Phi}}\frac{\sin\frac{\pi}{2}\omega_x}{\omega_x} - \frac{d}{\sqrt{\Phi}}\cos\frac{\pi}{2}\omega_x \\ \frac{\sqrt{\Phi}}{d}\cos\frac{\pi}{2}\omega_x & \frac{\sin\frac{\pi}{2}\omega_x}{\omega_x} \end{pmatrix} \times$$

$$\times \begin{pmatrix} \mathrm{Ch}\,2\beta\zeta & \frac{\mathrm{Sh}\,2\beta\zeta}{\beta\sqrt{\Phi}} \\ \sqrt{\Phi}\,\beta\,\mathrm{Sh}\,2\beta\zeta & \mathrm{Ch}\,2\beta\zeta \end{pmatrix} \times$$

$$\times \begin{pmatrix} \frac{\sin\frac{\pi}{2}\omega_x}{\omega_x} & -\frac{Z_o}{\sqrt{\Phi}}\frac{\sin\frac{\pi}{2}\omega_x}{\omega_x} - \frac{d}{\sqrt{\Phi}}\cos\frac{\pi}{2}\omega_x \\ \frac{\sqrt{\Phi}}{d}\cos\frac{\pi}{2}\omega_x & -\frac{Z_o}{d}\cos\frac{\pi}{2}\omega_x + \omega_x\sin\frac{\pi}{2}\omega_x \end{pmatrix}$$

in which $\beta^2 = D_0/4\Phi$ and $Z_c = z_c - \zeta$, $Z_o = z_o + \zeta$.

The cardinal elements are hence

$$\frac{1}{f_{xi}} = \frac{\mathrm{Ch}\,2\beta\zeta\,\sin\pi\,\omega_x}{d\,\omega_x} + \beta\,\mathrm{Sh}\,2\beta\zeta\left(\frac{\sin^2\frac{\pi}{2}\omega_x}{\omega_x^2} + \frac{\cos^2\frac{\pi}{2}\omega_x}{1-\omega_x^2}\right)$$

$$z_{Fi}^{(x)} = \zeta + f_{xi}\left(\mathrm{Ch}\,2\beta\zeta\,\cos\pi\,\omega_x + \mathrm{Sh}\,2\beta\zeta\,\mathrm{Sin}\,\pi\,\omega_x\frac{1-2\omega_x^2}{2\omega_x\sqrt{1-\omega_x^2}}\right)$$

$$z_{Hi}^{(x)} = z_{Fi}^{(x)} + f_{xi}\,.$$

Formulae for the cardinal elements of this model have been calculated by *Bernard* [41], which differ from those listed above in one particular: in $z_{Fi}^{(x)}$, *Bernard* has $\omega_x^2\sqrt{1-\omega_x^2}$ where I have $\omega_x\sqrt{1-\omega_x^2}$. Computations by *M. G. R. Thomson* (unpublished) appear to confirm the formulae listed above.

d) The triangular model. This idealized potential distribution is also more likely to be useful in calculating the properties of strong focusing lenses than those of short quadrupoles [196, 197]. The cardinal elements and transfer matrices are listed in [92], for the simplest case where no central plateau is present ($L = 0$ in Fig. 9b). In general, we have

$$D(z) = m(z+L) + D_0 \qquad \text{for} \quad z < -L$$
$$D(z) = D_0 \qquad \text{for} \quad |z| < L$$
$$D(z) = -m(z-L) + D_0 \quad \text{for} \quad z > L\,.$$

Substituting

$$u = 1 + \frac{m}{D_0}(z+L)\,, \quad \bar{u} = -u$$

$$v = -1 + \frac{m}{D_0}(z-L)\,, \quad \bar{v} = -v$$

we find that for $z < -L$, the equations of motion become

$$\frac{d^2x}{du^2} - t^3 u\, x = 0\,; \quad \frac{d^2y}{d\bar{u}^2} - t^3 \bar{u}\, y = 0$$

in which $t^3 = D_0^3/4m^2\Phi$, and for $z > L$,

$$\frac{d^2x}{d\bar{v}^2} - t^3 \bar{v}\, x = 0\,; \quad \frac{d^2y}{dv^2} - t^3 v\, y = 0\,.$$

Each of these equations has solutions in the form of Airy integral functions:

$$x(u) = \alpha\, \mathrm{Ai}(t\,u) + \beta\, \mathrm{Bi}(t\,u)$$
$$y(u) = \gamma\, \mathrm{Ai}(-t\,u) + \delta\, \mathrm{Bi}(-t\,u)$$

and

$$x(v) = \alpha^*\, \mathrm{Ai}(-t\,v) + \beta^*\, \mathrm{Bi}(-t\,v)$$
$$y(v) = \gamma^*\, \mathrm{Ai}(t\,v) + \delta^*\, \mathrm{Bi}(t\,v)\,.$$

The potential function D vanishes in the planes $z = \pm\zeta$, $\zeta = L + D_0/m$, and the transfer matrices between $-\zeta$ and ζ can be derived in the usual way, by joining the Airy functions to circular functions at the planes $z = \pm L$. For the special case $L = 0$, we find

$$\begin{pmatrix} x(\zeta) \\ p(\zeta) \end{pmatrix} = \begin{pmatrix} \xi_{11} & \xi_{12} \\ \xi_{21} & \xi_{22} \end{pmatrix} \begin{pmatrix} x(-\zeta) \\ y(-\zeta) \end{pmatrix}; \quad \begin{pmatrix} y(\zeta) \\ q(\zeta) \end{pmatrix} = \begin{pmatrix} \eta_{11} & \eta_{12} \\ \eta_{21} & \eta_{22} \end{pmatrix} \begin{pmatrix} y(-\zeta) \\ q(-\zeta) \end{pmatrix}$$

with

$$\xi_{11} = \xi_{22} = -\frac{\pi}{2\sqrt{3}}\{3(\mathrm{Ai}^2)'_t - (\mathrm{Bi}^2)'_t\}$$

$$\xi_{12} = -\frac{\lambda\pi}{\mu t\sqrt{\Phi}}\left\{(\mathrm{Ai\,Bi})'_t - \frac{\sqrt{3}}{2}(\mathrm{Ai}^2)'_t - \frac{1}{2\sqrt{3}}(\mathrm{Bi}^2)'_t\right\}$$

$$\xi_{21} = \frac{\mu t\pi\sqrt{\Phi}}{\lambda}\left\{(\mathrm{Ai\,Bi})'_t + \frac{\sqrt{3}}{2}(\mathrm{Ai}^2)'_t + \frac{1}{2\sqrt{3}}(\mathrm{Bi}^2)'_t\right\}$$

$$\eta_{11} = \eta_{22} = -\frac{\pi}{2\sqrt{3}}\{3(\mathrm{Ai}^2)'_{-t} - (\mathrm{Bi}^2)'_{-t}\}$$

$$\eta_{12} = \frac{\lambda\pi}{\mu t\sqrt{\Phi}}\left\{(\mathrm{Ai\,Bi})'_{-t} - \frac{\sqrt{3}}{2}(\mathrm{Ai}^2)'_{-t} - \frac{1}{2\sqrt{3}}(\mathrm{Bi}^2)'_{-t}\right.$$

$$\eta_{21} = -\frac{\mu t\pi\sqrt{\Phi}}{\lambda}\left\{(\mathrm{Ai\,Bi})'_{-t} + \frac{\sqrt{3}}{2}(\mathrm{Ai}^2)'_{-t} + \frac{1}{2\sqrt{3}}(\mathrm{Bi}^2)'_{-t}\right\}$$

where we have written

$$\mathrm{Ai}(0) = \lambda \quad \text{so that} \quad \mathrm{Bi}(0) = \sqrt{3}\,\lambda\,;$$
$$\text{and } \mathrm{Ai}'(0) = -\mu \text{ so that } \mathrm{Bi}'(0) = \sqrt{3}\,\mu.$$

The cardinal elements are hence

$$z_{Fi}^{(x)} = L + \frac{D_0}{m}\left(1 - \frac{\sqrt{\Phi}\,\xi_{11}}{\xi_{21}}\right); \quad z_{Fi}^{(y)} = L + \frac{D_0}{m}\left(1 - \frac{\sqrt{\Phi}\,\eta_{11}}{\eta_{21}}\right)$$

$$f_{xi} = \frac{D_0}{m}\frac{\sqrt{\Phi}}{\xi_{21}}; \quad f_{yi} = \frac{D_0}{m}\frac{\sqrt{\Phi}}{\eta_{21}}$$

$$z_{Hi}^{(x)} = L + \frac{D_0}{m}\left(1 + \frac{\sqrt{\Phi}(1 - \xi_{11})}{\xi_{21}}\right); \quad z_{Hi}^{(y)} = L + \frac{D_0}{m}\left(1 + \frac{\sqrt{\Phi}(1 - \eta_{11})}{\eta_{21}}\right).$$

By evaluating the triple matrix product which represents linear decay of D ,$D =$ constant and linear climb of D, the matrix elements and cardinal elements can easily be determined for the general case $L \neq 0$ whether or not the slopes on either side of the central plateau are equal in magnitude. Cf. *Sacerdoti* and *Uccelli* [154a, 227a].

e) Practical measurements and calculations. A number of measurements have been made [171–174, 183, 79–82] of the potential distribution in long quadrupoles; one of the most useful results of these involves the "equivalent length", L, defined by

$$L = \frac{1}{D_0} \int_{-\infty}^{\infty} D(z)\,\mathrm{d}z \quad \text{or} \quad L = \frac{1}{Q_0} \int_{-\infty}^{\infty} Q(z)\,\mathrm{d}z\,.$$

If this length is used instead of the real length of a long quadrupole, the formulae for the cardinal elements using the rectangular model prove to be very satisfactory; if the actual length is l and the bore-radius is a, it is found that $L = l + c\,a$, in which c remains close to 1.14 (that is, to $\pi - 2$ [152]). *Deltrap* [39] also uses this equivalent length for short lenses; his measurements have shown that the Glaser bell-shaped field is an excellent model, and that for $l = 9.9$ mm and $10 < a < 20$ mm, $L = l + (1.15 \pm 0.05)\,a$; (for the bell-shaped model, $L = \pi\,d/2$). The paraxial properties of individual electrostatic and magnetic quadrupoles have been measured by *Reisman* [152, 153], who shows that if the effective length is used, the rectangular model is satisfactory for the range of excitations he considered, which are all quite weak.

The potential distribution created by three sets of electrode shapes, planes tangent to a circle of bore-radius a, circular arcs separated by breaks of angle ε, and hyperbolae tangent to a circle, have been calculated by *Bernard* [11*, 13]; for long electrodes, he writes

$$\frac{\varphi(x, y, 0)}{\Phi_L} = K_1 \frac{x^2 - y^2}{a^2} + K_2 \frac{x^6 - y^6 - 15 x^2 y^2 (x^2 - y^2)}{a^6} + \cdots$$

and finds that for hyperbolae, $K_1 = 1$ and $K_2 = 0$, for planes $K_1 = 1.037$ and $K_2 = 0.009$, and for circular arcs $K_1 = 1.273\ \sin 2\varepsilon/2\varepsilon$ and $K_2 = 0.042 \sin 6\varepsilon/6\varepsilon$. For electrodes of length l, $D(z) = 4\Phi_L\,k_1(z)/a^2$ in *Bernard's* notation, and $k_1(z)$ is given by the following expression for circular arcs:

$$k_1(z) = -\frac{a^2}{\pi^2} \int_0^{\infty} \frac{\sin k\,z\,(1 - \cos k\,l)}{J_2(i\,k\,a)}\,k\,\mathrm{d}k$$
$$= \frac{1}{\pi}\left\{-2\mathfrak{Z}\left(\frac{z}{a}\right) + \mathfrak{Z}\left(\frac{z+l}{a}\right) + \mathfrak{Z}\left(\frac{z-l}{a}\right)\right\}$$

in which

$$\mathfrak{Z}(u) = \frac{1}{2\pi} \int_0^{\infty} \frac{k \sin k\,u}{J_2(i\,k)}\,\mathrm{d}k = -\left(2 + \frac{1}{2} \sum_{n=1}^{\infty} \frac{x_n\,\mathrm{e}^{-x_n u}}{J_1(x_n)}\right)$$

with $J_2(x_n) = 0$.

* The corresponding trajectories and cardinal elements are to be found in [12].

A large number of calculations of potential distributions are included in *Strashkevich*'s monograph [206] and articles [146, 148, 201, 205, 207, 208, 209, 211] in which the relativistic [202, 205, 207] and reduced [201] equations of motion are derived; the "point-pole" approximation has been used by *Archard* [3–7 *passim*]. The theory and practice of magnetic quadrupole field measurement are described in [61] and [37], and the effect of pole-piece shape is further discussed in [100, 101, 125, 147, 191, 194]. Precision measurements are described in [75], in which an electrolytic tank is employed, in [162], where a Hall probe is used, and in [221], with relaxation and a computer; *Koltay* and *Szabó* discuss asymmetrically fed quadrupoles in which opposite poles are no longer at the same potential [120–122]. Electrostatic lenses for which $l = 2a$ and $l = 8a$ have been investigated by *Orr* [141, 142].

Further information is to be found in [8, 8a, 26, 44, 103, 118, 128, 144, 195, 226, 227, 228a]; in [67], a method of matching a given transfer matrix to a quadrupole and drift spaces is described.

In *Blewett* [18], values of the focal length and the position of the focus are tabulated for values of βL between 0.1 and 1.6, using the rectangular model.

4.3. Aperture aberration coefficients

Of the geometrical aberrations, the aperture aberrations have been studied in the greatest detail; for the rectangular and bell-shaped models, the integrals in terms of which these coefficients are expressed can be evaluated explicitly, and a considerable amount of information about the behaviour of practical quadrupoles is also available. In this section, we consider first the formulae for the aperture aberrations in various cases, and then review the experimental work on quadrupole aperture aberration coefficients.

The aperture aberrations are given by

$$x_c^{(3)} = (30)\, x_a^3 + (12)\, x_a y_a^2 ,$$
$$y_c^{(3)} = (03)\, y_a^3 + (21)\, x_a^2 y_a ;$$

or

$$x_c^{(3)} = (30)^*\, p_o^3 + (12)^*\, p_o q_o^2 ,$$
$$y_c^{(3)} = (03)^*\, q_o^3 + (21)^*\, p_o^2 q_o ;$$

formulae for the coefficients $(p\, q) = (0\, 0\, p\, q)$ are given in § 3, together with instructions for deducing the coefficients $(p\, q)^*$.

The behaviour of these coefficients has only recently been studied methodically although, as we shall see, certain of them have been measured and calculated for various situations for more than a decade. We consider first the various analytical approaches, using the rectangular, bell-shaped and triangular models, and then examine the measurements recorded experimentally.

a) The rectangular model. This model is most appropriate when the purpose of the quadrupole lenses is strong focusing, and it is probably for this reason that formulae for even the aperture aberrations are given explicitly only by *Reisman* [152], *Yagi* [234, 235] and *Dymnikov, Fishkova* and *Yavor* [56, 60]. The equations of motion in the form (3.1) are employed by *Yagi*, who calculates the discontinuities of slope across the ends of the rectangle to which the discontinuities in D or Q give rise, and the aberrations associated with the zone over which D or Q is constant. It is convenient to set the origin of the z-coordinate at the beginning of the rectangle ($z = -\zeta$ in the notation of § 4.1). We write $2\zeta = L$, and the paraxial solutions become

$$x = x_0 \operatorname{Ch} \beta z + \frac{x_0'}{\beta} \operatorname{Sh} \beta z$$

$$y = y_0 \cos \beta z + \frac{y_0'}{\beta} \sin \beta z$$

within the rectangle. To the third-order approximation, the trajectories in the same region are written by *Yagi* thus:

$$x(z) = \{x_0 - f(0)\} \operatorname{Ch} \beta z + \frac{x_0' - \beta^2 x_0 y_0 y_0' - f'(0)}{\beta} \operatorname{Sh} \beta z + f(z)$$

$$y(z) = \{y_0 - g(0)\} \cos \beta z + \frac{y_0' + \beta^2 x_0 x_0' y_0 - g'(0)}{\beta} \sin \beta z + g(z)$$

in which $f(z)$ and $g(z)$ denote the following functions:

$$\begin{aligned}
128 \beta f(z) = {} & 3\{(\beta x_0 + x_0')^3 e^{3\beta z} + (\beta x_0 - x_0')^3 e^{-3\beta z}\} + \\
& + 4\beta z\{2(\beta^2 y_0^2 + y_0'^2) - 3(\beta^2 x_0^2 - x_0'^2)\}\{(\beta x_0 + x_0') e^{\beta z} + (\beta x_0 - x_0') e^{-\beta z}\} - \\
& - 2 e^{\beta z}(\beta^2 y_0^2 - y_0'^2 + 4\beta y_0 y_0')(\beta x_0 + x_0')(\sin 2\beta z - \cos 2\beta z) + \\
& - 4 e^{\beta z}(\beta^2 y_0^2 - y_0'^2 - \beta y_0 y_0')(\beta x_0 + x_0')(\sin 2\beta z + \cos 2\beta z) + \\
& + 2 e^{-\beta z}(\beta^2 y_0^2 - y_0'^2 + 4\beta y_0 y_0')(\beta x_0 - x_0')(\sin 2\beta z + \cos 2\beta z) - \\
& - 4 e^{-\beta z}(\beta^2 y_0^2 - y_0'^2 + \beta y_0 y_0')(\beta x_0 - x_0')(-\sin 2\beta z + \cos 2\beta z)
\end{aligned}$$

and

$$\begin{aligned}
128 \beta g(z) = {} & -6\beta y_0(\beta^2 y_0^2 - 3y_0'^2) \cos 3\beta z - 6y_0'(3\beta^2 y_0^2 - y_0'^2) \sin 3\beta z - \\
& - 8\beta z\{3(\beta^2 y_0^2 + y_0'^2) - 2(\beta^2 x_0^2 - x_0'^2)\}(\beta y_0 \sin \beta z - y_0' \cos \beta z) - \\
& - 2 e^{2\beta z}(\beta y_0 - 2y_0')(\beta x_0 + x_0')^2 (\sin \beta z + \cos \beta z) - \\
& - 2 e^{2\beta z}(2\beta y_0 + y_0')(\beta x_0 + x_0')^2 (\sin \beta z - \cos \beta z) + \\
& + 2 e^{-2\beta z}(\beta y_0 + 2y_0')(\beta x_0 - x_0')^2 (\sin \beta z - \cos \beta z) + \\
& + 2 e^{-2\beta z}(2\beta y_0 - y_0')(\beta x_0 - x_0')^2 (\sin \beta z + \cos \beta z)
\end{aligned}$$

[the quantities x_0 and y_0 denote positions in the plane $z = 0$. These are of course related to the positions in the object plane, $z = -z_o$, by the simple formulae $x_0 = x(z_o) + x_0' z_o$; $y_0 = y(z_o) + y_0' z_o$].

In a current plane $z_c = L + c$, therefore, we have

$$x(z_c) = x(L) + c\, x_L'$$

$$y(z_c) = y(L) + c\, y_L'$$

in which

$$x'_L = \beta x_0 \operatorname{Sh}\beta L + x'_0 \operatorname{Ch}\beta L + \Gamma_x$$

$$y'_L = -\beta y_0 \sin\beta L + y'_0 \cos\beta L + \Gamma_y$$

so that

$$x(z_c) = x_0(\operatorname{Ch}\beta L + \beta c \operatorname{Sh}\beta L) + \frac{x'_0}{\beta}(\operatorname{Sh}\beta L + \beta c \operatorname{Ch}\beta L) + \Delta_x$$

and

$$y(z_c) = y_0(\cos\beta L - \beta c \sin\beta L) + \frac{y'_0}{\beta}(\sin\beta L + \beta c \cos\beta L) + \Delta_y .$$

The terms Γ_x, Γ_y, Δ_x, Δ_y are defined by

$$\begin{aligned}\Gamma_x = &-\beta f(0) \operatorname{Sh}\beta L - f'(0) \operatorname{Ch}\beta L + f'(L) - \\ &-\beta^2 x_0 y_0 y'_0 \operatorname{Ch}\beta L + \{(\beta x_0 + x'_0)\, e^{\beta L} + (\beta x_0 - x'_0)\, e^{-\beta L}\} \times \\ &\times \{{}^1/_2\, \beta y y' \cos 2\beta L - {}^1/_4\, (\beta^2 y_0^2 - y_0'^2) \sin 2\beta L\}\end{aligned}$$

$$\begin{aligned}\Gamma_y = &\;\beta g(0) \sin\beta L - g'(0) \cos\beta L + g'(L) + \beta^2 x_0 x'_0 y_0 \cos\beta L - \\ &- {}^1/_4 \{(\beta x_0 + x'_0)^2 e^{2\beta L} - (\beta x_0 - x'_0)^2 e^{-2\beta L}\} \times \\ &\times (\beta y_0 \cos\beta L + y'_0 \sin\beta L)\end{aligned}$$

$$\Delta_x = c\,\Gamma_x - f(0) \operatorname{Ch}\beta L - \frac{f'(0) \operatorname{Sh}\beta L}{\beta} + f(L) - \beta x_0 y_0 y'_0 \operatorname{Sh}\beta L$$

$$\Delta_y = c\,\Gamma_y - g(0) \cos\beta L - \frac{g'(0) \sin\beta L}{\beta} + g(L) + \beta x_0 x'_0 y_0 \sin\beta L .$$

From these formulae and the relations between x_0, y_0 and $x(z_o)$, $y(z_o)$, we could extract all the aberration coefficients; the aberrations associated with an axial point object are listed in detail by *Yagi* [234, 235] but *Dymnikov* et al. [60c] have noticed that *Yagi*'s formulae are incomplete, and their calculations show that the effect of the missing terms is considerable. The aperture aberration coefficients are listed in [56, 60]; they are reproduced in § 5 because , although this is no limitation on their generality, they are given in a form suitable for combined electric and magnetic quadrupoles, producing coincident rectangles – they are thus particularly convenient for calculating the aberration coefficients of achromatic quadrupoles. *Strashkevich* too discusses these aberrations [212, 213].

b) The bell-shaped model. For this model too, the aberration coefficients can all be given in closed form, in principle at least, since the integrals involved can be evaluated in terms of circular or hyperbolic functions. This has been performed by *Glaser* [73], for the aperture aberration coefficients in the image plane of a stigmatic system. *Glaser* considers a system without an aperture, and retains the relativistic terms in ε. He has

$$-x_i^{(3)}/s_{xi} = (30)\, x_o'^3 + (12)\, x'_o y_o'^2 ; \quad -y_i^{(3)}/s_{yi} = (03)\, y_o'^3 + (21)\, x_o'^2 y'_o$$

in which*

$$(30) = \int_0^i \left\{ \frac{2}{\Phi} \frac{1+2\varepsilon\Phi}{1+\varepsilon\Phi} D_1 - 4\bar{\eta} Q_1 - \frac{1}{12\Phi^2} \frac{1+\frac{5}{2}\varepsilon\Phi(1+\varepsilon\Phi)}{(1+\varepsilon\Phi)^2} D^2 - \right.$$

$$- {}^1\!/_2\, \bar{\eta}^2 Q^2 + \frac{\bar{\eta}}{3\Phi} \frac{1+2\varepsilon\Phi}{1+\varepsilon\Phi} DQ + \frac{1}{96\Phi} \frac{1+2\varepsilon\Phi}{1+\varepsilon\Phi} D'' -$$

$$\left. - \frac{\bar{\eta}}{24} Q'' \right\} T_x^4 \, dz$$

$(30) \to (03)$ if $D \to -D$, $Q \to -Q$ and $T_x \to T_y$

$$(21) = (12) = \int_0^i \left\{ \left(\frac{1}{32} \frac{D^2}{\Phi^2(1+\varepsilon\Phi)^2} - \bar{\eta}^2 Q^2 + \frac{\bar{\eta}}{4\Phi} \frac{1+2\varepsilon\Phi}{1+\varepsilon\Phi} DQ - \right.\right.$$

$$\left. - \frac{6}{\Phi} \frac{1+2\varepsilon\Phi}{1+\varepsilon\Phi} D_1 + 12\bar{\eta} Q_1 \right) T_x^2 T_y^2 +$$

$$\left. + \frac{3}{4} \left(\frac{1}{4\Phi} \frac{1+2\varepsilon\Phi}{1+\varepsilon\Phi} D - \bar{\eta} Q \right) (T_x^2 T_y'^2 - T_x'^2 T_y^2) \right\} dz$$

and (30) and (03) can be transformed into

$$(30) = \int_0^i \left\{ \frac{2}{\Phi} \frac{1+2\varepsilon\Phi}{1+\varepsilon\Phi} D_1 - 4\bar{\eta} Q_1 - \frac{7}{96} \frac{1+\frac{16}{7}\varepsilon\Phi(1+\varepsilon\Phi)}{\Phi^2(1+\varepsilon\Phi)^2} D^2 + \right.$$

$$\left. + \frac{\bar{\eta}}{4\Phi} \frac{1+2\varepsilon\Phi}{1+\varepsilon\Phi} DQ - \frac{\bar{\eta}^2}{3} Q^2 \right\} T_x^4 dz +$$

$$+ \frac{1}{2} \int_0^i \left(\frac{1}{4\Phi} \frac{1+2\varepsilon\Phi}{1+\varepsilon\Phi} D - \bar{\eta} Q \right) T_x^2 T_x'^2 \, dz$$

$$(03) = \int_0^i \left\{ \frac{2}{\Phi} \frac{1+2\varepsilon\Phi}{1+\varepsilon\Phi} D_1 - 4\bar{\eta} Q_1 - \frac{7}{96} \frac{1+\frac{16}{7}\varepsilon\Phi(1+\varepsilon\Phi)}{\Phi^2(1+\varepsilon\Phi)^2} D^2 + \right.$$

$$\left. + \frac{\bar{\eta}}{4\Phi} \frac{1+2\varepsilon\Phi}{1+\varepsilon\Phi} DQ - \frac{\bar{\eta}^2}{3} Q^2 \right\} T_y^4 -$$

$$- \frac{1}{2} \int_0^i \left(\frac{1}{4\Phi} \frac{1+2\varepsilon\Phi}{1+\varepsilon\Phi} D - \bar{\eta} Q \right) T_y^2 T_y'^2 \, dz \, .$$

It is convenient to collect the octopole functions into a separate group,

$$\Psi = \frac{2}{\Phi} \frac{1+2\varepsilon\Phi}{1+\varepsilon\Phi} D_1 - 4\bar{\eta} Q_1$$

* We write $x = x_0 s_x + x_0' T_x$, $y = y_0 s_y + y_0' T_y$ so that $t = T/\sqrt{\Phi}$; the asterisks on the $(\alpha\,\beta)$ have been dropped. Since Φ is constant, we follow *Glaser* in incorporating the accelerating voltage into η; we write $\bar{\eta} = \eta/\sqrt{\Phi(1+\varepsilon\Phi)}$.

so that

$$(30) = \int_0^i \Psi\, T_x^4\, dz - q_x\,; \quad (03) = \int_0^i \Psi\, T_y^4\, \mathrm{d}z - q_y$$

$$(21) = (12) = -3 \int_0^i \Psi\, T_x^2\, T_y^2\, dz + 3q\,.$$

(q_x, q_y and q are obtainable by inspection.)

Writing

$$\frac{\varkappa_0}{d} = \frac{D_0^2}{12\Phi^2}\,\frac{1 + \frac{5}{2}\varepsilon\Phi(1+\varepsilon\Phi)}{(1+\varepsilon\Phi)^2} - \frac{\bar\eta}{3}\,\frac{D_0 Q_0}{\Phi}\,\frac{1+2\varepsilon\Phi}{1+\varepsilon\Phi} + \frac{\bar\eta^2}{2} Q_0^2$$

$$= 4k^4/3d^4 \text{ (electrostatic) or } k^4/2d^4 \text{ (magnetic)}$$

$$\varkappa_1 = \frac{1}{6d}\left(\frac{D_0}{4\Phi}\,\frac{1+2\varepsilon\Phi}{1+\varepsilon\Phi} - \bar\eta\, Q_0\right) = \frac{k^2}{6d^3}$$

$$\frac{\varkappa_2}{d} = \frac{D_0^2}{96\Phi^2}\,\frac{1}{(1+\varepsilon\Phi)^2} + \frac{\bar\eta}{12}\,\frac{D_0 Q_0}{\Phi}\,\frac{1+2\varepsilon\Phi}{1+\varepsilon\Phi} - \frac{\bar\eta^2}{3} Q_0^2$$

$$= k^4/6d^4 \text{ (electrostatic) or } -k^4/3d^4 \text{ (magnetic)},$$

Glaser finds

$$q_x = \int_0^\pi \{(\varkappa_0 + 6\varkappa_1)\sin^2\varphi - 5\varkappa_1\}\,(A\,\mathrm{Sh}\,\sigma_x\varphi + B\,\mathrm{Ch}\sigma_x\varphi)^4\,\mathrm{d}\varphi + Q_x$$

$$q_y = \int_0^\pi \{(\varkappa_0 - 6\varkappa_1)\sin^2\varphi + 5\varkappa_1\}\,(\bar A\sin\omega_y\varphi + \bar B\cos\omega_y\varphi)^4\,\mathrm{d}\varphi + Q_y$$

$$q = (\varkappa_2 - 3k^2\varkappa_1)\int_0^\pi (A\,\mathrm{Sh}\sigma_x\varphi + B\,\mathrm{Ch}\sigma_x\varphi)^2 \times$$

$$\times (\bar A\sin\omega_y\varphi + \bar B\cos\omega_y\varphi)^2\sin^2\varphi\,\mathrm{d}\varphi +$$

$$+ 3\varkappa_1\int_0^\pi \{\omega_y^2(A\,\mathrm{Sh}\,\sigma_x\varphi + B\,\mathrm{Ch}\sigma_x\varphi)^2\,(\bar A\cos\omega_y\varphi - \bar B\sin\omega_y\varphi)^2 -$$

$$- \sigma_x^2(A\,\mathrm{Ch}\sigma_x\varphi + B\,\mathrm{Sh}\sigma_x\varphi)^2(\bar A\sin\omega_y\varphi + \bar B\cos\omega_y\varphi)^2\}\sin^2\varphi\,\mathrm{d}\varphi + \bar Q$$

in which

$$A = \frac{z_0}{\sigma_x}\mathrm{Ch}\sigma_x\pi + \mathrm{Sh}\sigma_x\pi\,; \qquad B = -\frac{z_0}{\sigma_x}\mathrm{Sh}\sigma_x\pi - \mathrm{Ch}\sigma_x\pi$$

$$\bar A = \frac{z_0}{\omega_y}\cos\omega_y\pi - \sin\omega_y\pi\,; \quad \bar B = -\frac{z_0}{\omega_y}\sin\omega_y\pi - \cos\omega_y\pi\,.$$

The terms Q_x, Q_y and $\bar Q$ denote contributions from the second component of the doublet projector lens that *Glaser* is analysing; if the doublet separation is c, *Glaser* finds $(q_y - Q_y) \to Q_x$ if $\bar A \to a$, $\bar B \to b$; $(q_x - Q_x) \to Q_y$ if $A \to \bar a$, $B \to \bar b$; and $(q - \bar Q) \to \bar Q$ if $\bar A \to a$, $\bar B \to b$, $A \to \bar a$, $B \to \bar b$, in which

$$a = -(\sigma_x/\omega_y)A\cos\omega_y\pi - B(\sin\omega_y\pi + c\cos\omega_y\pi/\omega_y)$$

$$b = (\sigma_x/\omega_y)A\sin\omega_y\pi - B(\cos\omega_y\pi - c\sin\omega_y\pi/\omega_y)$$

$$\bar a = -(\omega_y/\sigma_x)\bar A\,\mathrm{Ch}\sigma_x\pi + \bar B(\mathrm{Sh}\sigma_x\pi - c\,\mathrm{Ch}\sigma_x\pi/\sigma_x)$$

$$\bar b = (\omega_y/\sigma_x)\bar A\,\mathrm{Sh}\sigma_x\pi - \bar B(\mathrm{Ch}\sigma_x\pi - c\,\mathrm{Sh}\sigma_x\pi/\sigma_x)$$

(For stigmatic, orthomorphic imagery, $b = \bar b$ and $\sigma_x\bar a = \omega_y a$) .

Writing integrals of the form

$$\int_0^\pi (A\,\mathrm{Sh}\sigma_x\,\varphi + B\,\mathrm{Ch}\sigma_x\,\varphi)^2\,(\overline{B}\,\sin\omega_y\,\varphi + \overline{A}\,\cos\omega_y\,\varphi)^2\,\sin^2\varphi\,\mathrm{d}\varphi$$

as

$$J = K\int_0^\pi \mathrm{Sh}^2(\sigma_x\,\varphi - \alpha)\,\sin^2(\omega_y\,\varphi - \overline{\alpha})\,\sin^2\varphi\,\mathrm{d}\varphi$$

we obtain

$$\begin{aligned}8J/K = -\pi &+ \int_0^\pi \mathrm{Ch}\,2(\sigma_x\varphi - \alpha)\,\mathrm{d}\varphi + \int_0^\pi \cos 2(\omega_y\,\varphi - \overline{\alpha})\,\mathrm{d}\varphi - \\ &- \int_0^\pi \mathrm{Ch}\,2(\sigma_x\,\varphi - \alpha)\cos 2(\omega_y\,\varphi - \overline{\alpha})\,\mathrm{d}\varphi - \\ &- \int_0^\pi \mathrm{Ch}\,2(\sigma_x\,\varphi - \alpha)\cos 2\varphi\,\mathrm{d}\varphi + \int_0^\pi \cos 2\varphi\,\mathrm{d}\varphi - \\ &- \int_0^\pi \cos 2(\omega_y\varphi - \overline{\alpha})\cos 2\varphi\,\mathrm{d}\varphi + \\ &+ \int_0^\pi \mathrm{Ch}\,2(\sigma_x\,\varphi - \alpha)\cos 2(\omega_y\,\varphi - \overline{\alpha})\cos 2\varphi\,\mathrm{d}\varphi;\end{aligned}$$

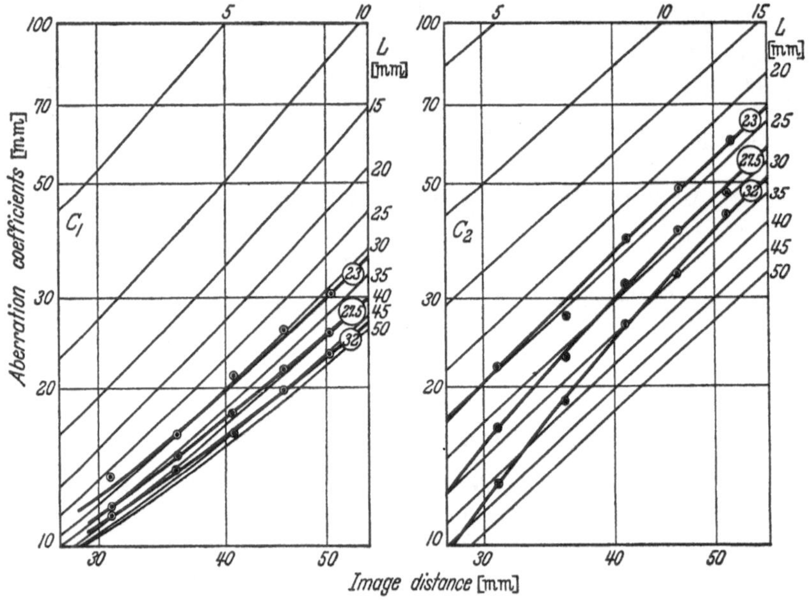

Fig. 14a. Measured and calculated values of the aberration coefficients as functions of the image distance (measured from the centre of the quadrupole) for several values of L, with an object distance of 21 cm

we also require

$$\begin{aligned}\overline{J} &= \int_0^\pi \mathrm{Ch}(p\,\varphi + q)\cos(r\,\varphi + s)\,\mathrm{d}\varphi \\ &= \frac{1}{p^2 + r^2}\left[r\,\mathrm{Ch}(p\,\varphi + q)\sin(r\,\varphi + s) + p\,\mathrm{Sh}(p\,\varphi + q)\cos(r\,\varphi + s)\right]_0^\pi.\end{aligned}$$

Formulae for this model have also been calculated by *Dymnikov, Fishkova* and *Yavor* [59], and as before, we defer discussion of them to § 5. *Tanguy* [219] considers the aberration coefficients in an "exit plane" as a function of the positions and slopes of rays at the "incident plane"; the incident and exit planes are equidistant from the centre of the lens, and for a parallel incident ray, Tanguy plots curves showing the distortion coefficients as functions of d and the excitation.

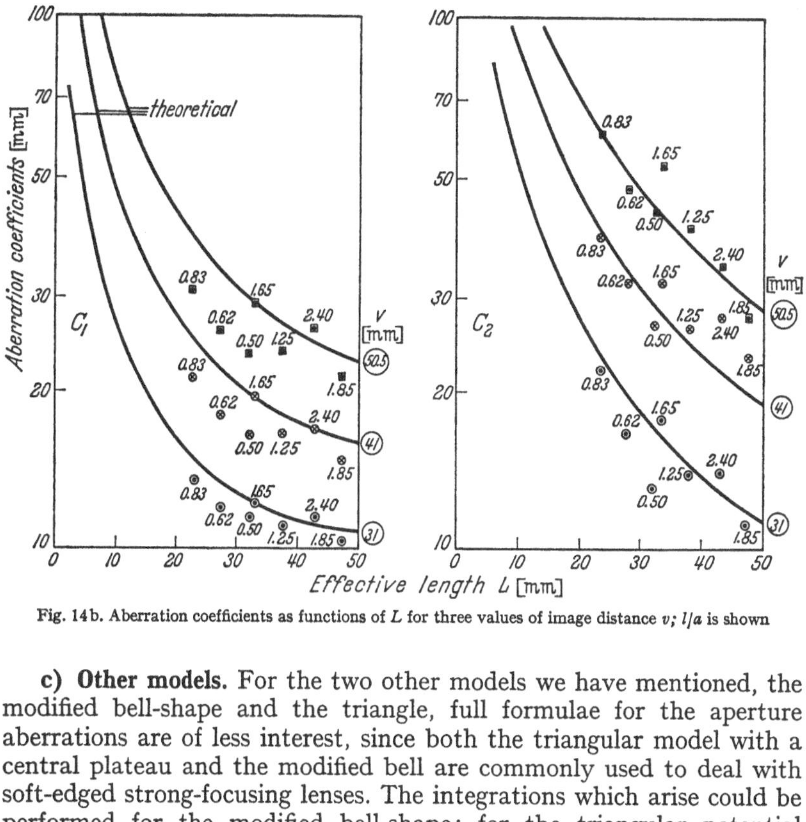

Fig. 14b. Aberration coefficients as functions of L for three values of image distance v; l/a is shown

c) Other models. For the two other models we have mentioned, the modified bell-shape and the triangle, full formulae for the aperture aberrations are of less interest, since both the triangular model with a central plateau and the modified bell are commonly used to deal with soft-edged strong-focusing lenses. The integrations which arise could be performed for the modified bell-shape; for the triangular potential (without a central plateau), the coefficients are listed in [92], but it does not seem possible to evaluate the integrals other than numerically.

d) Measurements. Several measurements of the aperture aberration coefficients of short (magnetic) quadrupole lenses have been made by *Deltrap* [38, 39]; writing $\Delta x = C_1 \alpha^3 + C_2 \alpha \beta^2$, $\Delta y = D_1 \beta^3 + D_2 \alpha^2 \beta$ at the real line-focus plane, in which α and β denote the ray slopes at this plane, he has measured and calculated C_1 and C_2 for a series of lenses, real length l mm, bore-radius a mm:

l:	9.9	9.9	9.9	20.0	20.0	29.6	29.6
a:	12.0	16.0	20.0	12.0	16.0	12.0	16.0

The object distance is 210 mm, and five values of "image-distance" are considered, between 31 and 52 mm; the latter is measured from the lens centre, and is nearly equal to the x-focal length. His results are shown in Figs. 14a and 14b, together with the calculated values. Further calculated values are illustrated in Figs. 14c, 14d, 14e and 14f, which

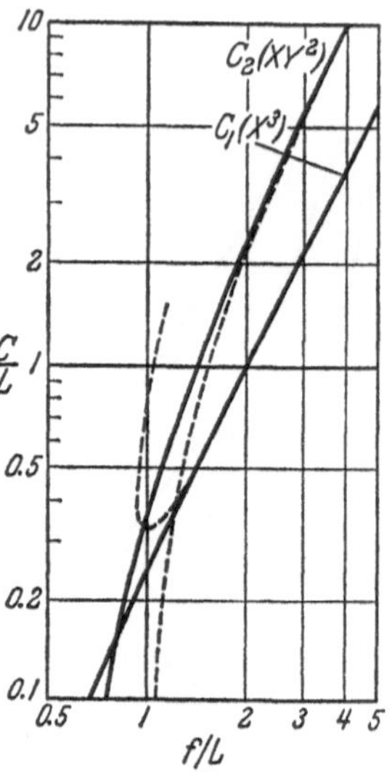

Fig. 14c

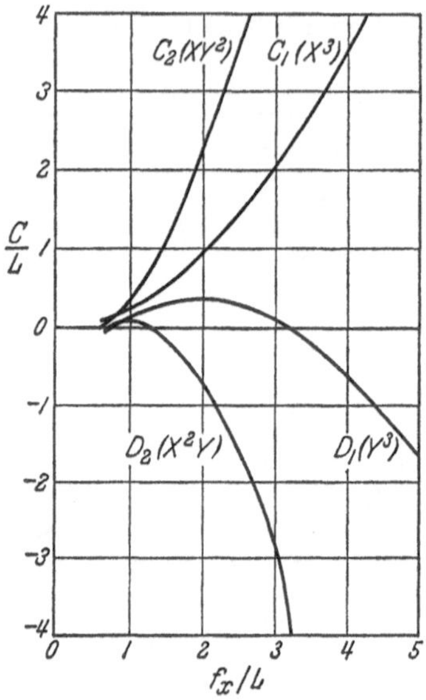

Fig. 14e

Fig. 14c. Aberration coefficients with a bell-shaped distribution for parallel incident rays. ——— values at cross-over, — — — asymptotic values

Fig. 14e. Aberration coefficients at a real cross-over (bell-shaped distribution)

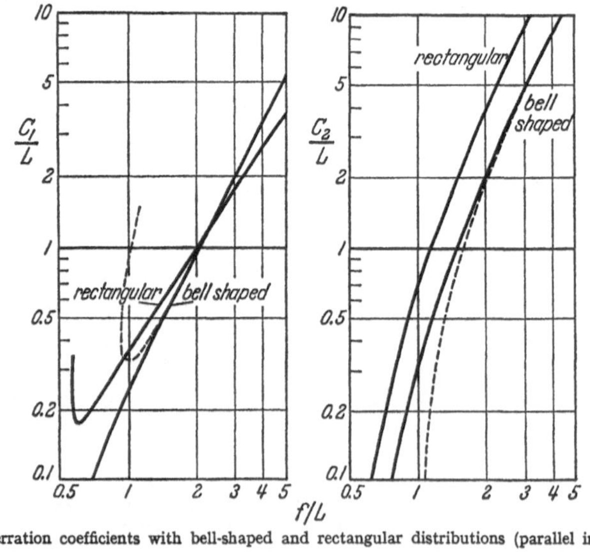

Fig. 14d. Aberration coefficients with bell-shaped and rectangular distributions (parallel incident beam)

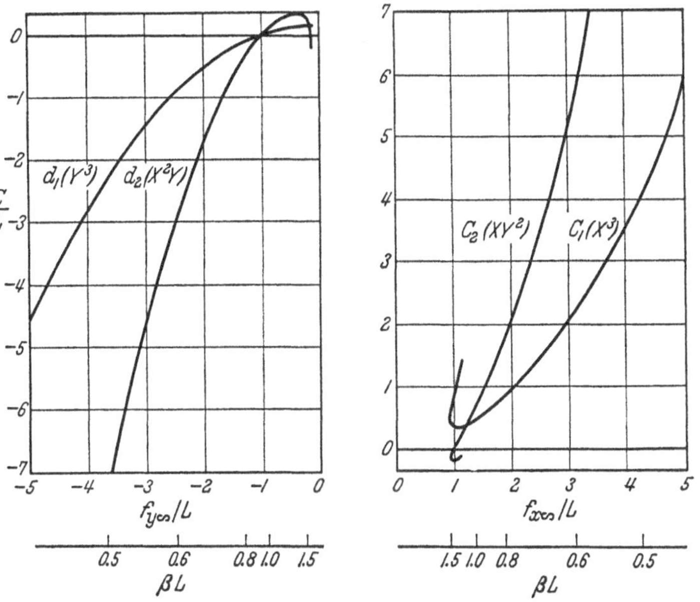

Fig. 14f. Asymptotic values of the aberration coefficients (bell-shaped distribution). (By courtesy of Dr. *Deltrap*)

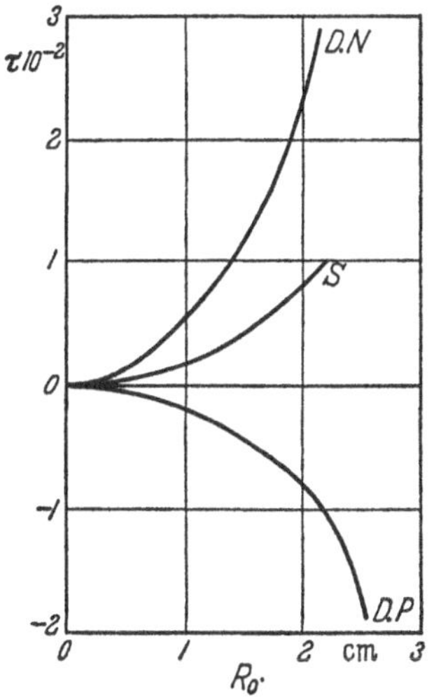

Fig. 15a

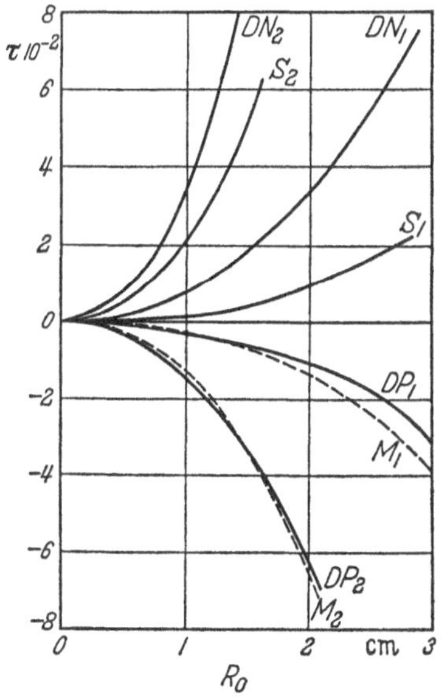

Fig. 15b

Fig. 15a. The aberration τ of a single quadrupole, as a function of the radius R_0 of the hollow cylindrical incident beam. 1. Symmetrical excitation: $\varphi(\pm a, 0, 0) = -\Phi_1$; $\varphi(0, \pm a, 0) = \Phi_1(S)$. 2. Positive unsymmetrical excitation: $\varphi(\pm a, 0, 0) = 0$; $\varphi(0, \pm a, 0) = 2\Phi_1(DP)$. 3. Negative unsymmetrical excitation: $\varphi(\pm a, 0, 0) = -2\Phi_1$; $\varphi(0, \pm a, 0) = 0(DN)$

Fig. 15b. The aberration, τ, of a symmetrical doublet. S_1, DP_1, DN_1: first line focus, S_2, DP_2, DN_2: second line focus, M_1, M_2: corresponding magnetic doublet. (By courtesy of Drs. *Septier* and *van Acker* and *Nuclear Instruments and Methods*) show clearly the difference between the actual aberrations at the line focus and the asymptotic aberration coefficients. Detailed discussion of the errors and validity of these values is to be found in [39].

Many measurements at the line foci of long magnetic quadrupoles have been made by *Septier* [175–179, see 81, 82], and at the line foci of electrostatic quadrupoles, symmetrically and asymmetrically excited, by *Septier* and *van Acker* [180, 181, 184, 185]. They find that with asymmetric excitation (which introduces round lens and octopole components into the potential expansion, φ_0- and φ_4-terms), the sign of the transverse aberration at the line focus of a single quadrupole and at the line foci of a doublet can be reversed (Figs. 15a,b); their measure of the aberration, τ, can be shown to be proportional to $\{(12) - (30)\}$ in the x-direction, and to $\{(21) - (03)\}$ in the y-direction [96]. Further measurements are recorded by *Orr* [141, 142], who investigated the behaviour of a laminated quadrupole, with which the potential distribution along the electrodes could be varied widely.

4.4. Lens systems

The search for combinations of quadrupole lenses that will perform certain specific tasks satisfactorily is still in active progress, and in this section we give a succinct account of the types of system that have been investigated, and indicate the general nature of the conclusions reached by the various investigators. Apart from strong focusing, quadrupole lenses are most likely to be used in conjunction with a round lens, to correct the primary spherical aberration of the latter; or alone, with a view to obtaining a lens with less spherical aberration than its round counterpart, but otherwise much the same kind of properties; or to produce a fine line image, broadened by very little aperture aberration [18a, 122c]; or to render anamorphotic images orthomorphic [64].

a) The quadrupole doublet. This, the simplest of quadrupole lens systems, has been the subject of repeated study. In *Glaser* [73], formulae and graphs for the cardinal elements of a doublet consisting of either two rectangular models or two bell-shaped distributions are given, together with formulae for the aperture aberrations of a bell-shaped doublet; *Dhuicq* [43] and *Septier* [175–179] and *van Acker* [180, 181, 184] supplement this, and *Yagi* gives formulae for the aberration coefficients associated with a point object on the axis in a rectangular (magnetic) doublet [234]; *Dymnikov* et al. have drawn attention to errors in *Yagi*'s formulae [60c].

(i) The rectangular doublet. If the two components are identical but inclined to one another at 90°, and each is of length L, the cardinal elements are given by

$$L/f_i = \beta^2 L D \sin\beta L \operatorname{Sh}\beta L + \beta L(\sin\beta L \operatorname{Ch}\beta L - \cos\beta L \operatorname{Sh}\beta L)$$
$$z_{F_i}^{(x)} = \bar{z} - L(\beta D \sin\beta L \operatorname{Ch}\beta L + \sin\beta L \operatorname{Sh}\beta L - \cos\beta L \operatorname{Ch}\beta L)$$
$$z_{F_i}^{(y)} = \bar{z} + L(\beta D \cos\beta L \operatorname{Sh}\beta L + \sin\beta L \operatorname{Sh}\beta L + \cos\beta L \operatorname{Ch}\beta L)$$

in which the distance between the rectangles is D and $\bar{z}$ is the coordinate of the end-face of the second lens. If the system is to be stigmatic for all points,

$$D = -\frac{2}{\beta}\frac{\operatorname{Sh}\beta L\sin\beta L}{\operatorname{Sh}\beta L\cos\beta L + \operatorname{Ch}\beta L\sin\beta L} = -\frac{2}{\beta}\frac{1}{\cot\beta L + \coth\beta L}$$

in which case

$$z_{Fi} = \bar{z} - \frac{1}{\beta}\frac{\sin\beta L\cos\beta L + \operatorname{Sh}\beta L\operatorname{Ch}\beta L}{\operatorname{Sh}^2\beta L - \sin^2\beta L}$$

$$f_i = -\frac{1}{\beta}\frac{\operatorname{Sh}\beta L\cos\beta L + \operatorname{Ch}\beta L\sin\beta L}{\operatorname{Sh}^2\beta L - \sin^2\beta L}$$

and the system is automatically orthomorphic; D is positive for certain ranges of values of βL, the first of which is $2.3652\ (\cong 3\pi/4) \leqq \beta L \leqq \pi$ but in this range, the focus always lies within the system. Fig. 16a illustrates the behaviour of f/L, $(z_{Fi}^{(x)} - \bar{z})/L$ and $(z_{Fi}^{(y)} - \bar{z})/L$ as functions of $x = \beta L$ for $D = 0$, and Fig. 16b shows $\lambda = D/L$, f/L and $(z_{Fi} - \bar{z})/L$ as functions of βL for a stigmatic system.

The properties of a doublet consisting of two separated or adjacent rectangles, equal in length but corresponding to different excitations, are discussed by *Enge* ([62]; cf. the chapter on cylindrical and quadrupole lenses in [111]); curves are plotted relating object and image distances, and giving the magnifications for adjacent rectangles, and rectangles separated by a distance equal to the width of either rectangle.

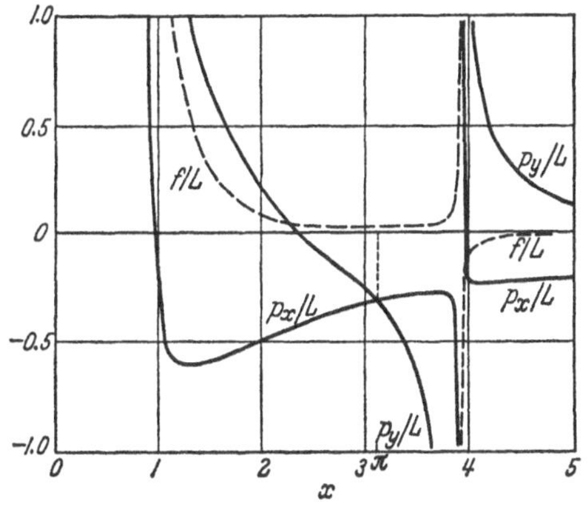

Fig. 16a. The focal length and position of the focus as functions of βL for $D = 0$

Extensive measurements of the cardinal elements and distortion coefficients of a magnetic doublet, intended for use as a projective lens, are to be found in *Reisman's* thesis [152, cf. 193].

A good account of doublets, regarded as strong-focusing devices, is to be found in [198]. Further comment is to be found in [2], [9] (doublet followed by a stigmator), [19, 20, 36a, 47, 49, 68, 112, 113, 123],

[124] in which formulae are given for the third order aberration coefficients of an electrostatic quadrupole doublet, [137, 154, 192, 220, 220a, 229, 231] and [230] in which formulae and experimental work are described.

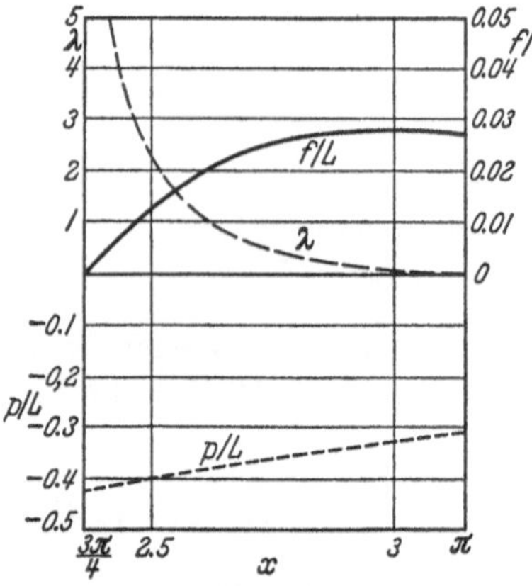

Fig. 16b. The separation D, the focal length and position of the focus for stigmatic imagery as functions of βL (the first range in which D is positive). (By courtesy of M. *Dhuicq*)

In *Blewett* [18], the positions of the principal planes and the focal lengths are tabulated for rectangles of equal length L in contact and separated by distances equal to $L/2$, L and $2L$; for each situation $0.1 \leq \beta_1 L \leq 1.6$ and $0.1 \leq \beta_2 L \leq 1.6$. *Lu* and *Carr* [124a] give graphs of operation for a doublet which focuses in one plane and behaves as a telescope in the other. *Möller, Dhuicq* and *Septier* [140a] discuss whether a doublet, a symmetric triplet or a symmetric quadruplet can provide a system having a positive or negative unit transfer matrix.

(ii) The bell-shaped doublet. This has been studied by *Glaser* [73, 74], and formulae for the case $\Phi(z) \neq$ constant are given by *Dušek* [45, 46]. The latter finds that the condition for stigmatic (and hence orthomorphic) imagery with two equal bells distance D apart is

$$\frac{2d}{D} = \frac{1}{k^2}\left(\omega_x \cot 2K\,\omega_x - \omega_y \cot 2K\,\omega_y\right)$$

or

$$\frac{2d}{D} = \frac{1}{k^2\sqrt{1-k_0^2}} \times \left(\Omega_x \cot 2K\,\Omega_x - \Omega_y \cot 2K\,\Omega_y\right)$$

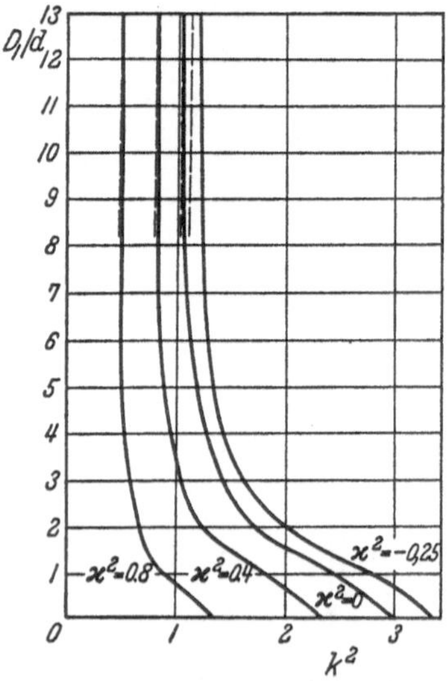

Fig. 17a

Fig. 17a—c. (a) The lens separation for stigmatic imagery with a bell-shaped doublet. (b) The corresponding focal length. (c) The distance of the image focus from the centre of the final quadrupole. (By courtesy of *Dušek* and *Optik* [46])

for negative or positive values of the round lens potential (see Fig. 17); $k_0 = \varkappa/\sqrt{1+\varkappa^2}$. The corresponding focal lengths and foci are given by

$$f_i = -\,d\,\omega_x\,\omega_y\,\frac{\omega_y \cos 2K\,\omega_y \sin 2K\,\omega_x - \omega_x \cos 2K\,\omega_x \sin 2K\,\omega_y}{\omega_y^2 \sin^2 2K\,\omega_x - \omega_x^2 \sin^2 2K\,\omega_y}$$

or

$$f_i = -d \frac{\Omega_x \Omega_y}{\sqrt{1-k_0^2}} \frac{\Omega_y \cos 2K\,\Omega_y \sin 2K\,\Omega_x - \Omega_x \cos 2K\,\Omega_x \sin 2K\,\Omega_y}{\Omega_y^2 \sin^2 2K\,\Omega_x - \Omega_x^2 \sin^2 2K\,\Omega_y}$$

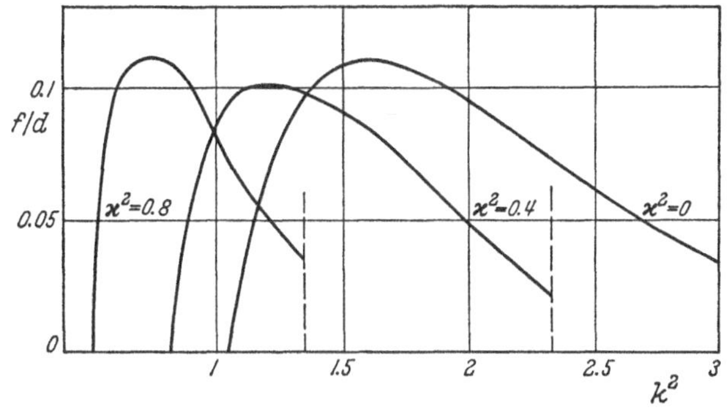

Fig. 17b

and

$$z_{Fi} = \frac{D}{2} + \frac{d}{2}\,\omega_x\,\omega_y \frac{\omega_y \sin 4K\,\omega_x - \omega_x \sin 4K\,\omega_y}{\omega_y^2 \sin^2 2K\,\omega_x - \omega_x^2 \sin^2 2K\,\omega_y}$$

or

$$z_{Fi} = \frac{D}{2} + \frac{d}{2} \frac{\Omega_x \Omega_y}{\sqrt{1-k_0^2}} \frac{\Omega_y \sin 4K\,\Omega_x - \Omega_x \sin 4K\,\Omega_y}{\Omega_y^2 \sin^2 2K\,\Omega_x - \Omega_x^2 \sin^2 2K\,\Omega_y}$$

respectively.

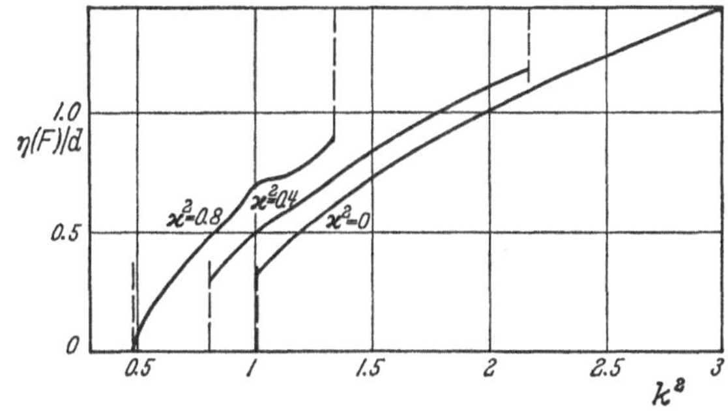

Fig. 17c

When $\Phi =$ constant, these formulae reduce to the expressions obtained by *Glaser*:

$$\frac{2d}{D} = \frac{\sqrt{k^2-1}}{k^2} \coth \pi \sqrt{k^2-1} - \frac{\sqrt{k^2+1}}{k^2} \cot \pi \sqrt{k^2+1}$$

$$= \frac{\sigma_x}{k^2} \coth \pi\,\sigma_x - \frac{\omega_y}{k^2} \cot \pi\,\omega_y$$

$$f_i = -d\,\omega_y\,\sigma_x \frac{\omega_y \operatorname{Sh} \pi\,\sigma_x \cos \pi\,\omega_y - \sigma_x \operatorname{Ch} \pi\,\sigma_x \sin \pi\,\omega_y}{\omega_y^2 \operatorname{Sh}^2 \pi\,\sigma_x - \sigma_x^2 \sin^2 \pi\,\omega_y}$$

$$z_{Fi} = \frac{D}{2} + \frac{d}{2}\,\omega_y\,\sigma_x \frac{\omega_y \operatorname{Sh} 2\pi\,\sigma_x - \sigma_x \sin 2\pi\,\omega_y}{\omega_y^2 \operatorname{Sh}^2 \pi\,\sigma_x - \sigma_x^2 \sin^2 \pi\,\omega_y}\,.$$

b) The quadrupole triplet. Only the rectangular model has been used to study this system, as in some cases [18, 63] the combination was to be used for strong focusing, while *Dhuicq*, although admittedly considering

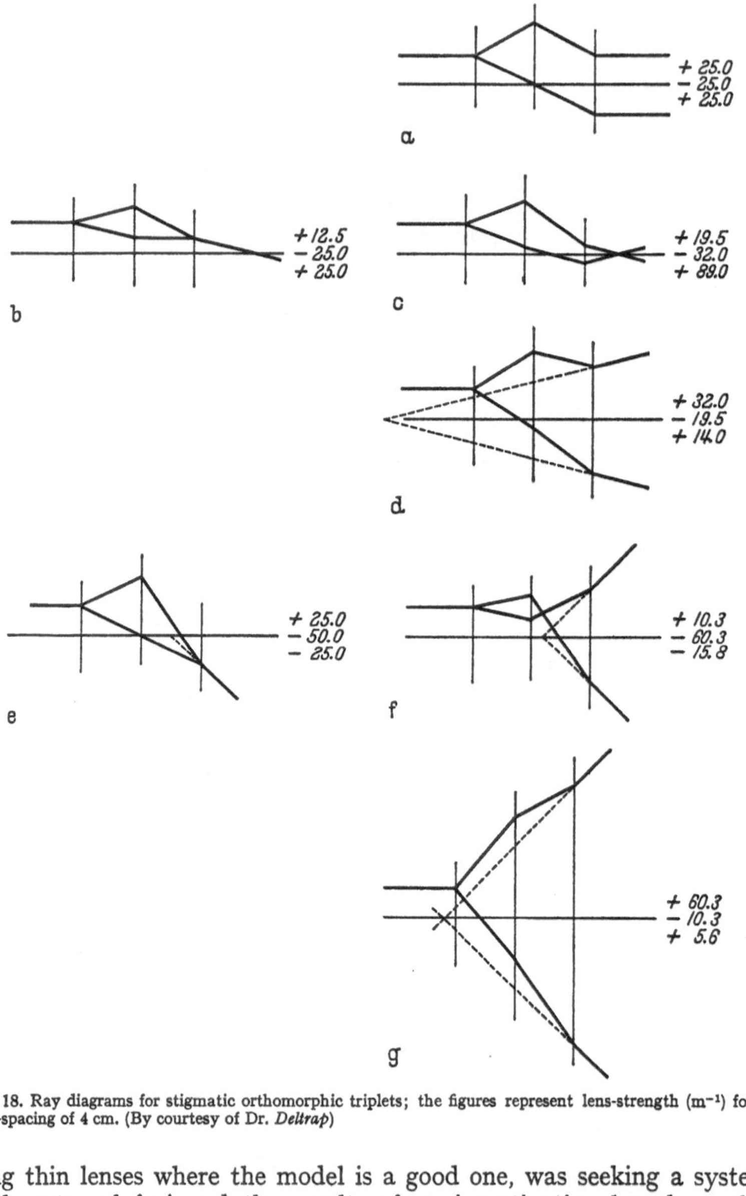

Fig. 18. Ray diagrams for stigmatic orthomorphic triplets; the figures represent lens-strength (m^{-1}) for a lens-spacing of 4 cm. (By courtesy of Dr. *Deltrap*)

long thin lenses where the model is a good one, was seeking a system with external foci and the results of an investigation based on this simple model could reasonably be expected to indicate whether or not such a system was attainable.

The paper by *Blewett* [18] contains tabulated values of focal lengths and abscissae of the principal plane for values of βL between 0.1 and 1.6 and for a range of geometries. If L is the effective length of the two outer lenses and L' is that of the central lens, and the spacing is λ, *Blewett* considers the cases $\lambda = 0$, $L = L'$; $\lambda = L/2$, $L = L'$; $\lambda = L$, $L = L'$; $\lambda = 2L$, $L = L'$; $\lambda = 0$, $L' = 2L$; $\lambda = L/2$, $L' = 2L$; $\lambda = L$, $L' = 2L$; and $\lambda = 2L$, $L' = 2L$. The design of stigmatic triplets is described in [17].

In *Enge* [63], the triplets considered are such that $L' = 2L$, and $\lambda = 0$ or $\lambda = L$. The excitations required to provide a particular image distance for a given object distance are plotted, together with the magnifications.

Dhuicq [43, 43a] gives formulae for the positions of the foci and the focal lengths for a symmetrical triplet with arbitrary (equal) separations between the components (λ), and arbitrary excitations of the central and outer lenses. For $\lambda = 0$, the conditions for stigmatic imagery are deduced, and for the values of λ/L used by *Blewett*, a graphical method is devised to yield these conditions. For this type of triplet, the focus remains immersed within the lens, however.

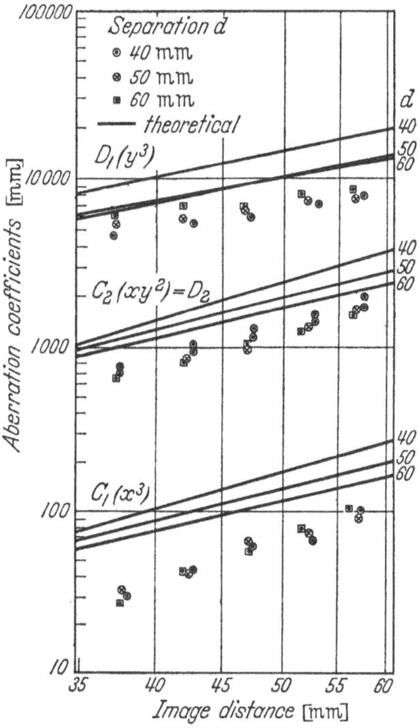

Fig. 19. Measurements of the aberration coefficients of stigmatic and orthomorphic triplets; inter-element distances: 30, 40 and 50 mm, corresponding to object distances 151, 131, 111 mm. (By courtesy of Dr. *Deltrap*)

All the work described above is concerned with symmetrical triplets, possessing a central plane of geometrical and electrical symmetry. *Deltrap* [39] abandons this restrictive arrangement and considers the possibility of obtaining stigmatic orthomorphic imagery with three differently excited quadrupoles; the aperture aberration coefficients for favourable arrangements are then measured. A range of possible systems is illustrated in Fig. 18, calculated on the thin lens approximation. The aberration coefficients are measured using a quadrupole of length 12.8 mm and bore-radius 16.6 mm ($L = 31$ mm); the (equal) lens separations were 40, 50 or 60 mm for which the object distances to the *centre* of the first quadrupole were 151, 131 or 111 mm. The results, together with calculated values, are illustrated in Fig. 19.

Quadrupole triplets are also mentioned in [99, 114, 222].

c) Systems of four or more quadrupole lenses. These systems fall into three classes: quadruplets with a high degree of symmetry, in which the lens excitations of the first and fourth lenses are the same, or equal and opposite, as are those of the second and third quadrupoles, and the two outer lens spacings are equal (studied by *Dymnikov*, *Fishkova* and *Yavor*, in Leningrad, and *Dhuicq*, in Paris); quadruplets in which the excitations are chosen independently, to satisfy some overall condition (*Deltrap* [39]); and sets of four elements, again with unrelated excitations, designed to produce a quadruplet with some specific properties [21, 22, 232, 233]. (*Burfoot's* work [21, 22] differs in such fundamental respects from all the other studies that it is convenient to classify it separately.) A five-quadrupole objective has been designed by *Bauer* [10], but because of the sensitivity of quadrupole systems to misalignments and other mechanical faults, there is little incentive to employ more than four or five elements for optical purposes; for strong focusing, however, channels with many quadrupoles are often convenient – the rules of matrix algebra lead to a number of laws governing their behaviour, which we shall not discuss here (see [198, 182, 154a] or the other review articles cited in § 1). Notwithstanding, the first electron optical system using non-rotationally symmetrical elements to counter spherical aberration did contain a large number of independent components, but cylindrical lenses were employed, and not quadrupoles [166, 139, 140].

(i) *Burfoot's* system [21, 22]. The system proposed by *Burfoot* was the result of an attempt to design a four-electrode electrostatic lens, free of first-order distortion and astigmatism, and free of aperture aberration. The lens consisted of an ordinary electrostatic einzel lens, suitably modified, followed by a quadrupole lens; the "suitable" modifications converted the first and third electrodes of the einzel lens into elements producing $\Phi(z)$ and $D(z)$ components, while the second electrode produced $\Phi(z)$ and $D_1(z)$.

The paraxial equations of motion,

$$x'' + \frac{1}{2}\frac{\Phi'}{\Phi}x' + \frac{\Phi'' - D}{4\Phi}x = 0\,, \quad y'' + \frac{1}{2}\frac{\Phi'}{\Phi}y + \frac{\Phi'' + D}{4\Phi}y = 0$$

are solved by *Burfoot* by transferring the terms in D to the right hand sides of the equations, and using a method of successive approximations: the first approximation is thus the solution of the round lens equations. *Burfoot* uses the hybrid ray-pair, $g(z)$ and $T(z)$, but the lower limits in his aberration formulae (all are z_0) require $s(z)$ and $T(z)$ [or $t(z)$]. No attempt is made here to alter *Burfoot's* analysis, as any re-examination of his work would involve much more radical alteration to take advantage of the simpler formulae for the coefficients which have since become available; furthermore different "peak-shapes" from those he used now seem more appropriate.

Following *Burfoot*, therefore, we set $x = x_0 g + x_a T/T_a$, $y = y_0 g + y_a T/T_a$ as a first approximation, and better approximations modify these to

$$x = x_0(g + N_1 + N_2 + \cdots) + x_a(T + M_1 + M_2 + \cdots)/T_a$$
$$y = y_0(g - N_1 + N_2 + \cdots) + y_a(T - M_1 + M_2 - \cdots)/T_a\,.$$

Variation of parameters yields

$$\sqrt{\Phi_o}\, M_1 = T \int_o^c D^* g\, T \,\mathrm{d}z - g \int_o^c D^* T^2 \,\mathrm{d}z$$

$$\sqrt{\Phi_o}\, N_1 = T \int_o^c D^* g^2 \,\mathrm{d}z - g \int_o^c D^* g\, T \,\mathrm{d}z$$

with $D^* = D/4\sqrt{\Phi}$; a recurrence formula gives M_{i+1}, N_{i+1} in terms of M_i, N_i respectively.

If the plane z_i is to be a stigmatic image plane conjugate to z_o, both $T + M_2 + M_4 + \cdots$ and $M_1 + M_3 + \cdots$ must vanish there; for orthomorphism, $M_1' + M_3' + \cdots$ must also be zero.

The aberration coefficients are calculated on the assumption that only M_1 and N_1 need be retained in the paraxial solutions, and that squares and higher powers of M_1 and N_1 may be neglected. Into the third-order equations of motion, *Burfoot* makes the following substitutions:

$$x'' y'^2 \to \bar{x}_a \bar{y}_a^2 (T'' T' - 2 T'' M_1' + T' M_1'')\, T'$$

$$x' y' y'' \to \bar{x}_a \bar{y}_a^2 (T'' - M_1'')\, T'^2$$

$$x'' x^2 \to \bar{x}_a^3 (T'' T + 2 T'' M_1 + T M_1'')\, T$$

$$x'' y^2 \to \bar{x}_a \bar{y}_a^2 (T'' T - 2 T'' M_1 + T M_1'')\, T$$

$$x^3 \to \bar{x}_a^3 (T + 3 M_1)\, T^2$$

$$x y^2 \to \bar{x}_a \bar{y}_a^2 (T - M_1)\, T^2$$

$$x y'^2 \to \bar{x}_a \bar{y}_a^2 (T T' - 2 T M_1' + T' M_1)\, T'$$

$$x' y y' \to \bar{x}_a \bar{y}_a^2 (T - M_1)\, T'^2$$

$$x x'^2 \to \bar{x}_a^3 (T T' + 2 T M_1' + T' M_1)\, T'$$

$$x' x^2 \to \bar{x}_a^3 (T T' + 2 T' M_1 + T M_1')\, T$$

$$x' y^2 \to \bar{x}_a \bar{y}_a^2 (T T' - 2 T' M_1 + T M_1')\, T$$

$$x'^3 \to \bar{x}_a^3 (T' + 3 M_1')\, T'^2$$

$$x' y'^2 \to \bar{x}_a \bar{y}_a^2 (T' - M_1')\, T'^2 .$$

($\bar{x}_a = x_a/T_a$, $\bar{y}_a = y_a/T_a$) so that the primary aperture aberration coefficients are given by

$$x^{(3)} = \frac{\bar{x}_a^3}{\sqrt{\Phi_o}} \left(T \int_o^i g \frac{W_a + W_c}{\sqrt{\Phi}} \,\mathrm{d}z - g \int_o^i T \frac{W_a + W_c}{\sqrt{\Phi}} \,\mathrm{d}z \right) + \frac{\bar{x}_a \bar{y}_a^2}{\sqrt{\Phi_o}} \left(T \int_o^i g \frac{W_b + W_d}{\sqrt{\Phi}} \,\mathrm{d}z - g \int_o^i T \frac{W_b + W_b}{\sqrt{\Phi}} \,\mathrm{d}z \right)$$

$$y^{(3)} = \frac{\bar{y}_a^3}{\sqrt{\Phi_o}} \left(T \int_o^i g \frac{W_a - W_c}{\sqrt{\Phi}} \,\mathrm{d}z - g \int_o^i T \frac{W_a - W_c}{\sqrt{\Phi}} \,\mathrm{d}z \right) + \frac{\bar{x}_a^2 \bar{y}_a}{\sqrt{\Phi_o}} \left(T \int_o^i g \frac{W_b - W_d}{\sqrt{\Phi}} \,\mathrm{d}z - g \int_o^i T \frac{W_b - W_d}{\sqrt{\Phi}} \,\mathrm{d}z \right)$$

in which

$$W_a = W_o + {}^1\!/_4\, D\,(T'^2 M_1 - 2\,T\,T''\,M_1 + 2\,T\,T'\,M_1' - T^2 M_1'') -$$
$$- {}^1\!/_4\, D'\,(T\,T'\,M_1 + {}^1\!/_2\, T^2 M_1') - {}^1\!/_2\, D''\,T^2 M_1 + 2 D_1 T^3$$
$$W_b = W_o + {}^1\!/_4\, D\,(T'^2 M_1 - 2\,T\,T''\,M_1 - 4\,T\,T'\,M_1' + T^2 M_1'') -$$
$$- {}^1\!/_4\, D'\,(T\,T'\,M_1 - {}^1\!/_2\, T^2 M_1') - 6 D_1 T^3$$
$$W_c = {}^3\!/_{32}\, \Phi^{(\mathrm{iv})}\, T^2 M_1 + {}^1\!/_8\, \Phi'''\,(2\,T\,T'\,M_1 + T^2 M_1') +$$
$$+ \Phi''\,(T\,T''\,M_1 - {}^1\!/_4\, T'^2 M_1 - {}^1\!/_2\, T\,T'\,M_1' + {}^1\!/_4\, T^2 M_1'') -$$
$$- {}^3\!/_2\, \Phi'\,T'^2 M_1' + {}^1\!/_4\, D\,(T\,T'^2 - T^2\,T'') - {}^1\!/_8\, D'\,T^2\,T' - {}^1\!/_{24}\, D''\,T^3 +$$
$$+ 6 D_1 T^2 M_1$$
$$W_d = - {}^1\!/_{32}\, \Phi^{(\mathrm{iv})}\, T^2 M_1 + {}^1\!/_8\, \Phi'''\,(-2\,T\,T'\,M_1 + T^2 M_1') +$$
$$+ \Phi''\,(-{}^1\!/_2\, T\,T''\,M_1 - {}^3\!/_4\, T'^2 M_1 + T\,T'\,M_1 + {}^1\!/_4\, T^2 M_1'') +$$
$$+ {}^1\!/_2\, \Phi'\,T'^2 M_1' + 2\Phi\,(T'\,T''\,M_1' - T'^2 M_1'') +$$
$$+ {}^1\!/_4\, D\,(3\,T\,T'^2 + T^2\,T'') + {}^1\!/_8\, D'\,T^2\,T' + 6 D_1 T^2 M_1$$

with

$$W_o = {}^1\!/_{32}\, \Phi^{(\mathrm{iv})}\, T^3 + {}^1\!/_8\, \Phi'''\,T^2\,T' + {}^1\!/_4\, \Phi''\,(T^2\,T'' - T\,T'^2) - {}^1\!/_2\, \Phi'\,T'^3 .$$

The system is thus free of aperture aberration if

$$T_i \int_o^i \frac{g W}{\sqrt{\Phi}}\,\mathrm{d}z - g_i \int_o^i \frac{T W}{\sqrt{\Phi}}\,\mathrm{d}z = 0$$

for all four expressions for W. The conditions which must be satisfied if a system is to be stigmatic, orthomorphic and free of aperture aberration then prove to be

$$I + a + \zeta - T_i(\bar{I} + \bar{a} + \bar{\zeta})/s_i + \chi_a = 0;$$
$$I + b - 3\zeta - T_i(\bar{I} + \bar{b} - 3\bar{\zeta})/s_i + \chi_b = 0;$$
$$c + \xi - T_i(\bar{c} + \bar{\xi})/s_i + \chi_c = 0;$$
$$d + \xi - T_i(\bar{d} + \bar{\xi})/s_i + \chi_d = 0;$$
$$e - T_i\,\bar{e}/s_i + \chi_e = 0; \quad f - T_i(\bar{f} + \sqrt{\Phi_o})/s_i + \chi_f = 0$$

in which

$$\zeta = 2 \int_{123} (D_1\, T^4/\sqrt{\Phi}\,)\,\mathrm{d}z; \qquad \bar{\zeta} = 2 \int_{123} (D_1\, g\, T^3/\sqrt{\Phi}\,)\,\mathrm{d}z$$

$$\xi = 6 \int_{123} (D_1\, T^3 M_1/\sqrt{\Phi}\,)\,\mathrm{d}z; \qquad \bar{\xi} = 6 \int_{123} (D_1\, g\, T^2 M_1/\sqrt{\Phi}\,)\,\mathrm{d}z$$

$$I = \int_{123} (T W_o/\sqrt{\Phi}\,)\,\mathrm{d}z; \qquad \bar{I} = \int_{123} (g W_o/\sqrt{\Phi}\,)\,\mathrm{d}z$$

$$f = \int_{123} (M_1\, T D/4\sqrt{\Phi}\,)\,\mathrm{d}z; \qquad \bar{f} = \int_{123} (M_1\, g\, D/4\sqrt{\Phi}\,)\,\mathrm{d}z$$

$$e = \int_{123} (T^2 D/4\sqrt{\Phi}\,)\,\mathrm{d}z; \qquad \bar{e} = \int_{123} (g\, T\, D/4\sqrt{\Phi}\,)\,\mathrm{d}z$$

$$a = \int_{123} (T\, W_{aD}/\sqrt{\Phi}\,)\,\mathrm{d}z; \qquad \bar{a} = \int_{123} (g\, W_{aD}/\sqrt{\Phi}\,)\,\mathrm{d}z$$

$$b = \int_{123} (T\, W_{bD}/\sqrt{\Phi}\,)\,\mathrm{d}z; \qquad \bar{b} = \int_{123} (g\, W_{bD}/\sqrt{\Phi}\,)\,\mathrm{d}z$$

$$c = \int_{123} (T\, \overline{W}_c/\sqrt{\Phi}\,)\,\mathrm{d}z; \qquad \bar{c} = \int_{123} (g\, \overline{W}_c/\sqrt{\Phi}\,)\,\mathrm{d}z$$

$$d = \int_{123} (T\, \overline{\overline{W}}_d/\sqrt{\Phi}\,)\,\mathrm{d}z; \qquad \bar{d} = \int_{123} (g\, \overline{\overline{W}}_d/\sqrt{\Phi}\,)\,\mathrm{d}z.$$

W_{aD}, W_{bD} denote the terms in W_a, W_b involving D alone, and $\overline{W}_c$, $\overline{W}_d$ denote the terms in W_c, W_d containing Φ and D. The integration (123) is taken over the first three electrodes, and between the third and fourth electrodes, a field-free region is assumed to exist. The quantities χ are all associated with the fourth electrode (or corrector) placed at $z = z_O$, and writing

$$C = {}^1/_4 \int_4 D \, \mathrm{d}z \, , \qquad S = {}^1/_4 \int_4 D (z - z_O)^2 \, \mathrm{d}z \, .$$

Burfoot finds [M' stands for $M_1'(z)$ in the field-free region between the third and fourth electrodes]:

$$\sqrt{\Phi_O}\, \chi_a = \frac{3}{4} T_i' M' (T_O^2 C + T_i'^2 S) - $$
$$ - \frac{T_i}{g_i} \left\{ T_O M' C \left(T_i' g_O - \frac{1}{4} T_O g_i' \right) + \frac{3}{4} T_i'^2 g_i' M' S \right\}$$
$$\sqrt{\Phi_O}\, \chi_b = - \frac{3}{4} T_i' M' (3 T_O^2 C + T_i'^2 S) - $$
$$ - \frac{T_i}{g_i} \left\{ - T_O M' C \left(2 T_i' g_O + \frac{1}{4} T_O g_i' \right) - \frac{3}{4} T_i'^2 g_i' M' S \right\}$$
$$\sqrt{\Phi_O}\, \chi_c = \frac{1}{2} T_i'^2 (T_O^2 C + T_i'^2 S) - $$
$$ - \frac{T_i}{g_i} \left\{ T_O T_i' \left(T_i' g_O - \frac{1}{2} T_O g_i' \right) C + \frac{1}{2} T_i'^3 g_i' S \right\}$$
$$\sqrt{\Phi_O}\, \chi_d = - \frac{1}{2} T_i'^2 (T_O^2 C + T_i'^2 S) - \frac{T_i}{g_i} \left\{ - \frac{1}{2} T_i' g_i' (T_O^2 C + T_i'^2 S) \right\}$$
$$\sqrt{\Phi_O}\, \chi_e = T_O^2 C + T_i'^2 S - \frac{T_i}{g_i} (T_O g_O C + T_i' g_i' S)$$
$$\sqrt{\Phi_O}\, \chi_f = \frac{1}{2} T_i' M' S - \frac{T_i}{2 g_i} g_i' M' S \, .$$

The fourth electrode is treated as a thin lens. The function $D(z)$ is assumed to be a linear combination of two standard peak-shapes, $\Omega_m(z)$ for the first electrode and $\Omega_n(z)$ for the third electrode, so that*

$${}^1/_4 \, D(z) = m \, \Omega_m(z) + n \, \Omega_n(z)$$

and hence c, d and e are all of the form

$$c = m \, c_m + n \, c_n$$

while a, b and f become

$$a = m^2 a_m + m \, n \, a_{mn} + n^2 a_n \, .$$

For a lens with high magnification, *Burfoot* finds

$$n = (c_m - d_m) \left[-4 I / \{ (3 a_m + b_m) (c_n - d_n)^2 - (3 a_{mn} + b_{mn}) \times \right.$$
$$\left. \times (c_m - d_m) (c_n - d_n) + (3 a_n + b_n) (c_m - q_m)^2 \} \right]^{1/2}$$
$$m = - n (c_n - d_n) / (c_m - d_m)$$
$$\zeta = - I - a - {}^3/_4 \, C \, T_O^2 \, T_i' \, M' / \sqrt{\Phi_O}$$
$$\xi = - c - {}^1/_2 \, C \, T_O^2 \, T_i'^2 / \sqrt{\Phi_O}$$
$$z_O = z_b - s_b / (s_i' - T_i' \, \bar{e} / e)$$
$$C = - e \sqrt{\Phi_O} / T_O^2$$
$$z_i = z_b - s_b / \{ s_i' - T_i' (\bar{f} + \sqrt{\Phi_o}) / f \} \, .$$

The suffix b denotes the plane in which T vanishes.

---

* The factor ${}^1/_4$ is retained to avoid altering *Burfoot's* final formulae.

To obtain the peak shapes Ω_m and Ω_n, *Burfoot* writes *Laplace's* equation in oblate spheroidal coordinates (μ, ν, θ):

$$\frac{\partial}{\partial \mu}\left\{(\mu^2+1)\frac{\partial \varphi}{\partial \mu}\right\} + \frac{\partial}{\partial \nu}\left\{(1-\nu^2)\frac{\partial \varphi}{\partial \nu}\right\} + \\ + \frac{\partial}{\partial \theta}\left\{\frac{\nu^2+\mu^2}{(1+\mu^2)(1-\nu^2)}\frac{\partial \varphi}{\partial \theta}\right\} = 0$$

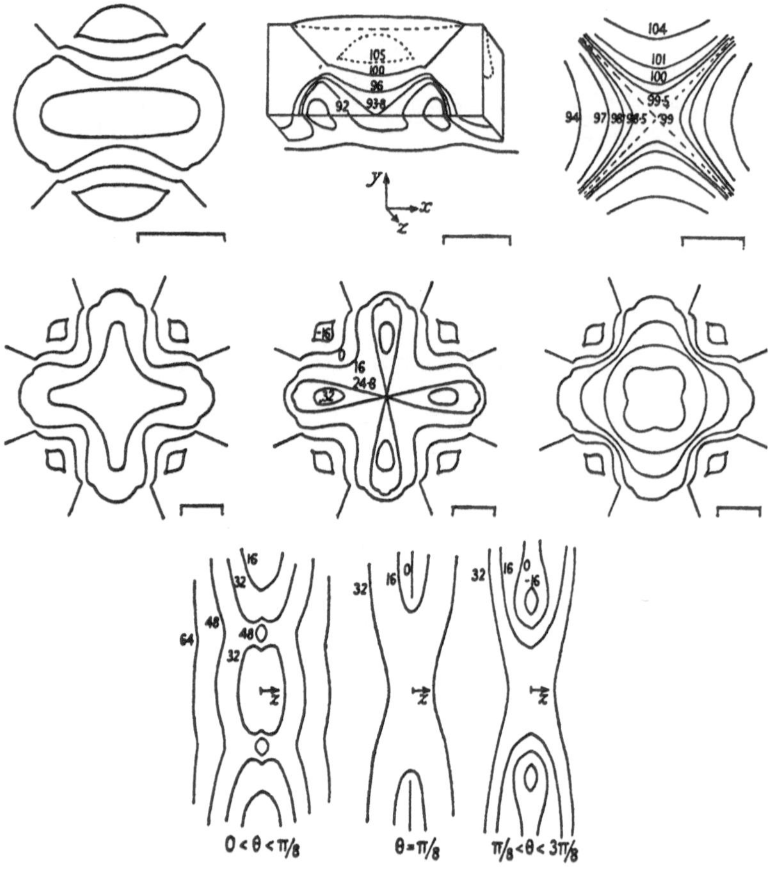

Fig. 20. *Burfoot's* lens. The first and third electrodes are shown Top Left or Top Centre; the fourth (corrector) electrode is a pure quadrupole (Top Right); possible forms of the second electrode are shown in the middle row; profiles corresponding to the central case are represented below. (By courtesy of Dr. *Burfoot*, the Institute of Physics and The Physical Society)

and finds

$$\varphi = \sum_{l=0}^{\infty} A_{0,l}\, P_l(\nu)\, Q_l(i\,\mu) + \cos 2\theta \sum_{l=2}^{\infty} A_{2,l}\, P_l^2(\nu)\, Q_l^2(i\,\mu) + \\ + \cos 4\theta \sum_{l=4}^{\infty} A_{4,l}\, P_l^4(\nu)\, Q_l^4(i\,\mu) + \cdots$$

in which P_l^m, Q_l^m are the associated Legendre functions of the first and second kinds;

using only the first terms of the summations,

$$\frac{1}{4} D(z) = \frac{A_{2,2}}{R^2} \left(\frac{P_2^2(\nu)}{1-\nu^2} \right)_{\nu=1} \cdot \frac{Q_2^2\,(iz/R)}{1+z^2/R^2}$$

$$D_1(z) = \frac{A_{4,4}}{R^4} \left(\frac{P_4^4(\nu)}{(1-\nu^2)^2} \right)_{\nu=1} \frac{Q_4^4\,(iz/R)}{(1+z^2/R^2)^2} \,.$$

Burfoot concludes that the shapes of the requisite electrodes are highly complex (Fig. 20) and that the tolerances on certain electrodes would in practice be of the order of some tens of Ångström units.

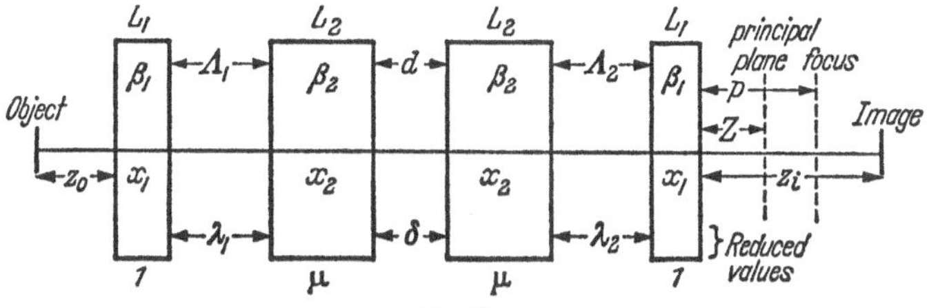

Fig. 21 a

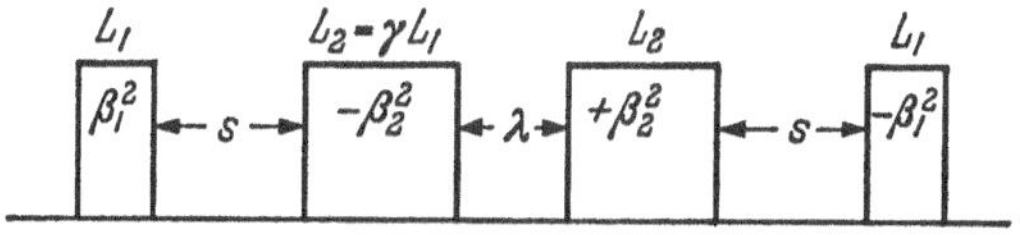

Fig. 21 b

Fig. 21 a and b. (a) The symmetric quadruplet. (b) The anti-symmetric quadruplet.

The work of *Whitmer* [232, 233] is related to that of *Burfoot*, but we shall not discuss it in detail here, as it is of considerably more tentative and exploratory a nature.

(ii) Quadruplets with high symmetry. Two types of quadruplet have been considered in detail, the symmetric quadruplet, by *Dhuicq* [41–43], and the antisymmetric arrangement, by a Russian group [48, 50, 54, 55, 57, 188, 236]; see Figs. 21 a and b, where the notation is explained. Strong focusing quadruplets are discussed in [32] and [164]; *Blewett's* quadruplets reduce to triplets [17].

The symmetric quadruplet. Dhuicq writes down formulae for the cardinal elements and the stigmatic condition, when all the lenses have the same length L; the expressions are highly complex, and *Dhuicq* therefore examines a number of special cases. For equal excitations on the lenses, $x_1 = x_2 = x = \beta L$, the condition for stigmatic imagery is the same as that for the symmetric doublet:

$$\lambda = -\frac{2}{x(\cot x + \coth x)} \,.$$

The focus lies within the lens.

For $\lambda = 0$, *Dhuicq* obtains x_1 as a function of x_2 for stigmatic imagery, in the range $0 \leqq x_1, x_2 \leqq 2\pi$; the separation δ and the cardinal elements are then plotted as functions of x_2, but the focus remains within the lens

(Figs. 21c–f). The cases $\mu \neq 1$, $\lambda = 0$ and $\mu \neq 1$, $\lambda = 0$ $x_1 = x_2 = x$ are briefly examined also.

A particular case, $\mu = 1$, $\lambda = 0.4$, $\delta = 0.27$ is studied in detail, theoretically and experimentally.

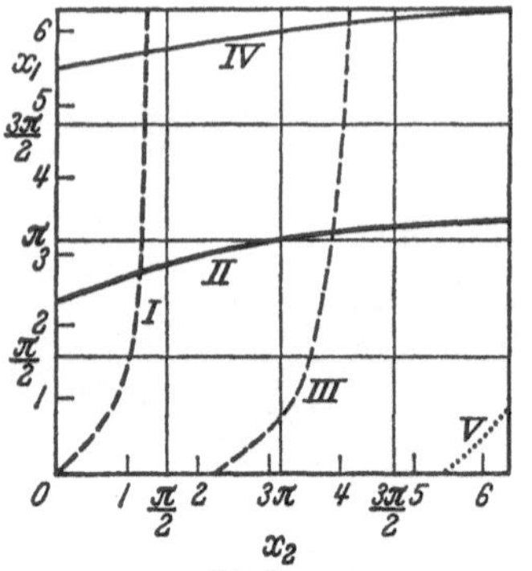

Fig. 21c

Fig. 21c—f. Two doublets of contiguous quadrupoles, all with the same equivalent length forming a quadruplet. (c) x_1 as a function of x_2 for stigmatic operation. (d) δ as a function of x_2. (e) f/L as a function of x_2. (f) p/L as a function of x_2. (By courtesy of M. *Dhuicq*).

The cardinal elements are first plotted as functions of x_1 and x_2 and the stigmatic working-point determined: $x_1 = 2.67$, $x_2 = 0.925$ which gives $p/L = -0.43$, $f/L = -0.05$, $Z/L = 0.38$. Experimentally, *Dhuicq* finds $x_1 = 2.70$, $x_2 = 0.93$, $p/L = -0.43$ and $f/L = -0.047$; he also gives a careful comparison of the theoretical and experimental curves showing the various quantities as functions of the excitations.

Finally, a quadruplet is examined, which is stigmatic only for one pair of conjugate points, and in which the rays from the axial object point leave the first doublet parallel to the axis (*King's* "focus-parallel" case [112]); if the first and second doublets are identical ($x_2 = x_3$ and $x_1 = x_4$, $\lambda_1 = \lambda_2$) the magnification will be unity. *Dhuicq* again examines

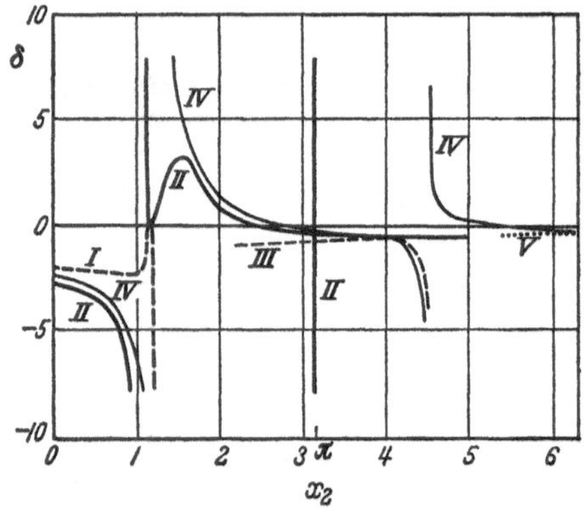

Fig. 21d

$x_1(x_2)$ and the reduced image distance*, $\xi(x_1)$ and $\xi(x_2)$, and a similar system is examined experimentally.

The antisymmetric quadruplet. This arrangement is mentioned by *Dhuicq* [43], who has recently studied it further** [43b], but it has been

* ξ denotes the distance from the end-face of the terminal quadrupole to the image plane, divided by L.

** In [43c] and [43d], *Dhuicq* and *Septier* give details of the paraxial and third order properties of such a quadruplet, equipped with octopoles to correct the aperture aberrations.

studied very throughly by *Dymnikov*, *Fishkova* and *Yavor* [48, 50, 54, 55]*; their work is summarized in [57] to which the reader is referred for a full discussion – here we merely reproduce their curves with little

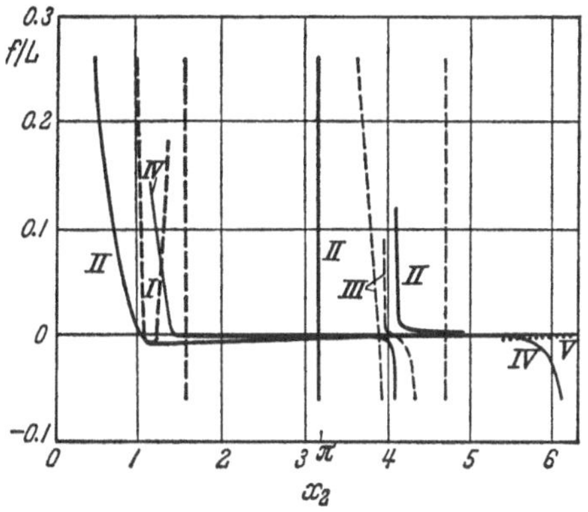

Fig. 21 e

comment. Their notation is shown in Fig. 21b. If the quadruplet is to have the same cardinal elements in the $x-z$ and $y-z$ planes, the

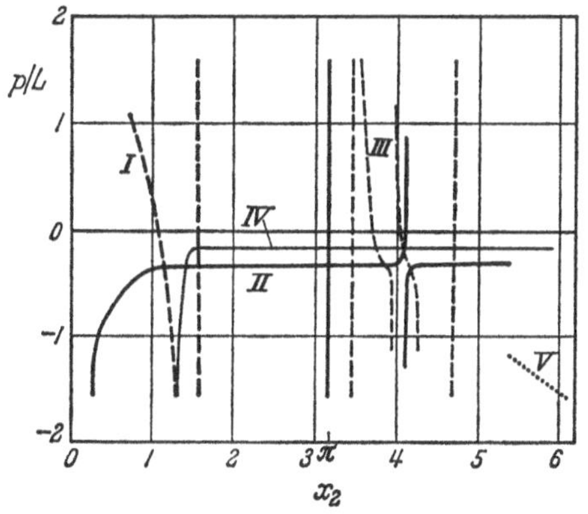

Fig. 21 f

excitations and geometrical parameters will be related:

$$\beta_1 L_1 = f(\beta_2 L_2, \gamma, s, \lambda) .$$

This function is referred to as the "load characteristic" of the system. For lenses of equal length ($\gamma = 1$), the load characteristic is plotted in Fig. 22. (Notice that in any system possessing symmetry about a centre plane except that the second half is rotated through 90° with respect to the first half – a symmetric doublet, for example, or an antisymmetric quadruplet – the x- and y-focal lengths are equal.)

* Work on this combination is also in progress at Cornell University, under the direction of *B. M. Siegel* (private communication).

Region I: $0 \leqq \beta_2 L_2 \leqq 2.3$, $0 \leqq \beta_1 L_1 \leqq 4.6$. The focal length f and the position of the focus, measured from the end of the last lens, z_F, are plotted in Fig. 23 for $s = 0$, and for a range of geometries; the latter

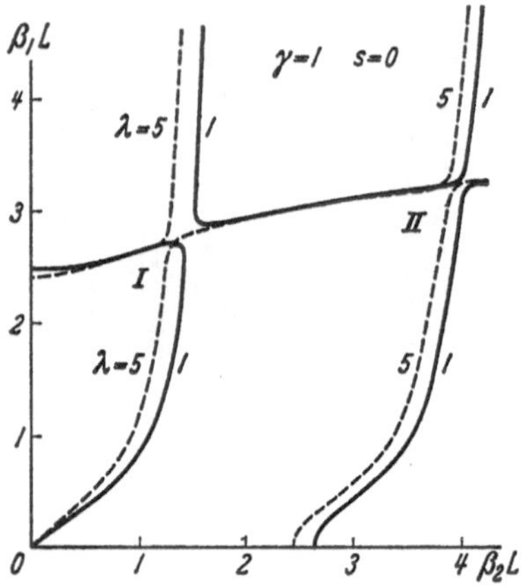

Fig. 22. The "load characteristic" for stigmatic orthomorphic imagery with a quadruplet. (Figs. 22—32 are taken from [57] with grateful acknowledgement)

are divided into categories in Fig. 24: region A corresponds to Fig. 23a and regions B and C to Fig. 23b. If the focus lies at the end of the last

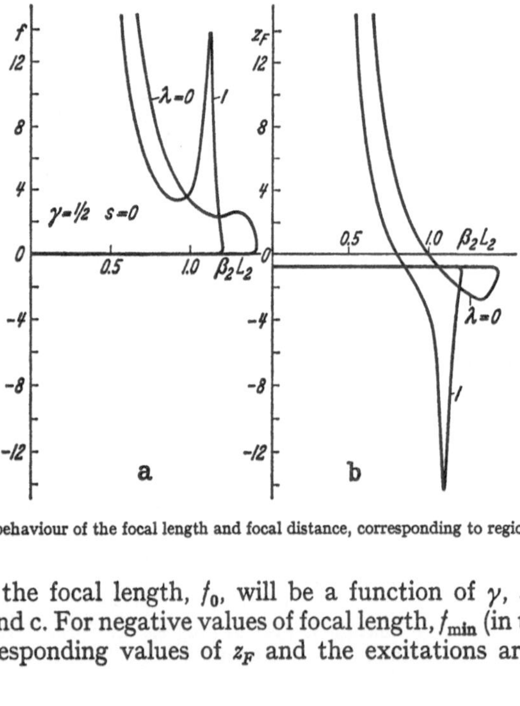

Fig. 23a. The behaviour of the focal length and focal distance, corresponding to region A of Fig. 24

lens, $z_F = 0$, the focal length, f_0, will be a function of γ, s and λ, see Figs. 25a,b and c. For negative values of focal length, $f_{\min}$ (in the region B) and the corresponding values of z_F and the excitations are plotted in

Fig. 26. Typical trajectories are illustrated in Fig. 27. Curves corresponding to region C are shown in Fig. 29.

Region II: $0 \leqq \beta_1 L_1 \leqq 6$; $2.3 \leqq \beta_2 L_2 \leqq 5$. For $\gamma = 1$, $s = 0$, $\lambda = 1$, the focal length and z_F are plotted as functions of $\beta_1 L_1$ in Fig. 28a

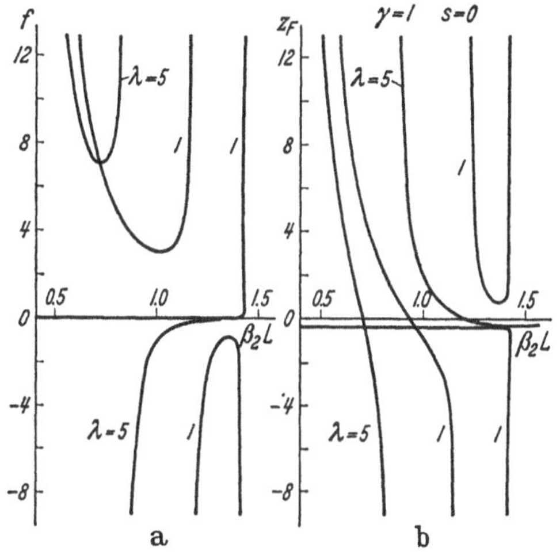

Fig. 23b. The behaviour of the focal length and focal distance corresponding to regions B and C of Fig. 24

(third branch) and in Fig. 28b (fourth branch); on the third branch, there are two intervals in which the system may be convergent, for one of

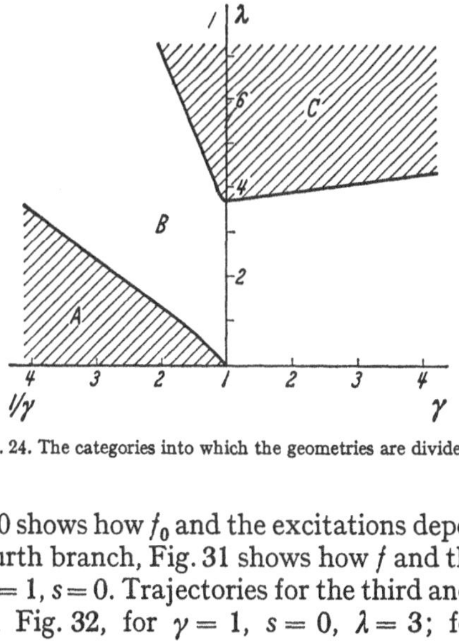

Fig. 24. The categories into which the geometries are divided

which $f > 0$. Fig. 30 shows how f_0 and the excitations depend upon the geometry. For the fourth branch, Fig. 31 shows how f and the excitations depend upon λ for $\gamma = 1$, $s = 0$. Trajectories for the third and fourth branches are illustrated in Fig. 32, for $\gamma = 1$, $s = 0$, $\lambda = 3$; for the former, f

$= 0.0096$, $z_F = 0.0490$, $\beta_1 L = 2.450$ and $\beta_2 L = 3.800$, and for the latter $f = -0.050$, $z_F = 0.001$, $\beta_1 L = 5.430$ and $\beta_2 L = 4.136$.

(iii) *Deltrap's* quadruplets [39, 40]. In his chapter on "Compound systems of quadrupoles", *Deltrap* shows that the range of types of stigmatic

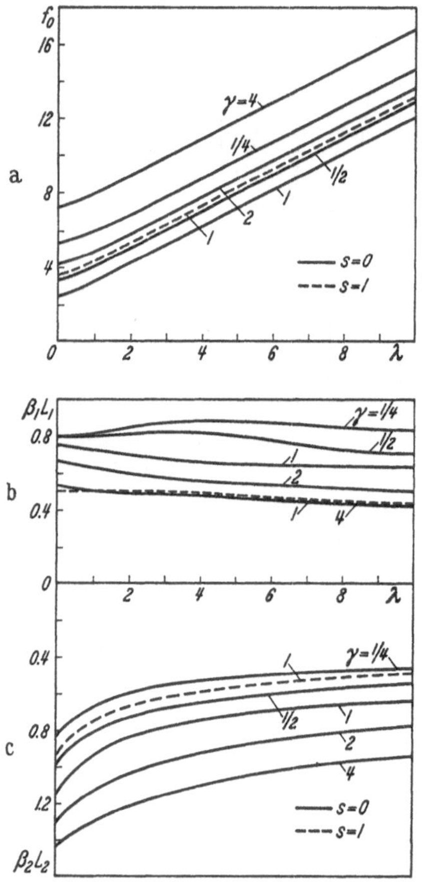

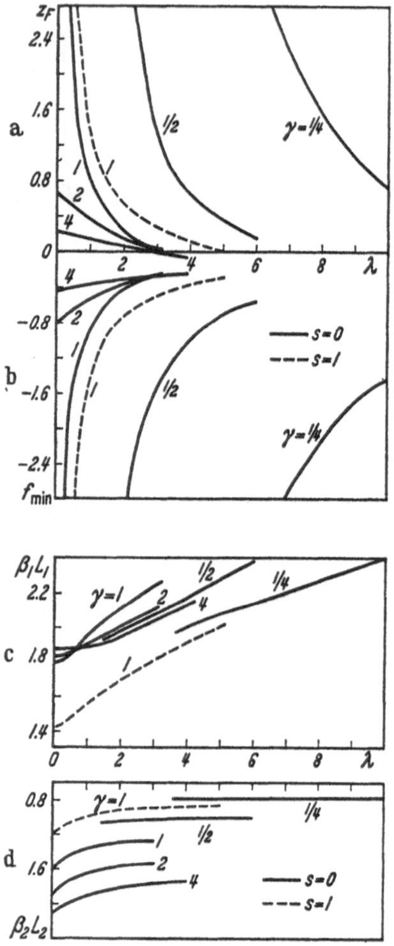

Fig. 25 Fig. 26

Fig. 25a—c. The parameters corresponding to (a) Focal length; (b) and (c): Excitations

Fig. 26a—d. Negative focal length. (a): Focal distance; (b) minimum focal length; (c) and (d): excitations

orthomorphic quadruplets is very wide, and illustrates schematically some fifty arrangements. The purpose of his research was to design a quadrupole-octopole system to correct the spherical aberration of a round lens, however, and it is interesting that the system he finally selected as the combination best fitted to do this possessed one of the

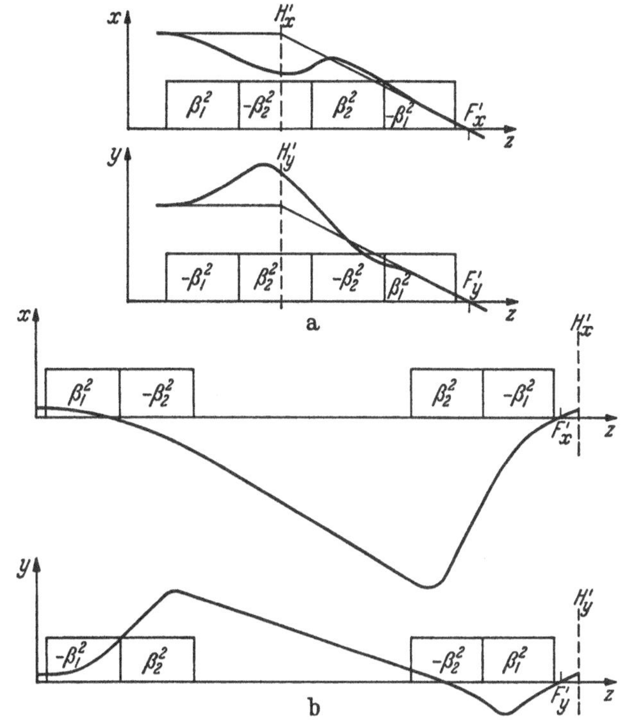

Fig. 27a and b. Typical trajectories in a quadruplet. (a) $\gamma = 1$, $s = 0$, $\lambda = 0$, focal length $= 2.530$, focal distance $= 0.143$, $\beta_1 L = 0.750$ and $\beta_2 L = 1.150$. (b) $\gamma = 1$, $s = 0$, $\lambda = 3$, focal length $= -0.269$, focal distance $= 0.056$, $\beta_1 L = 2.240$ and $\beta_2 L = 1.270$

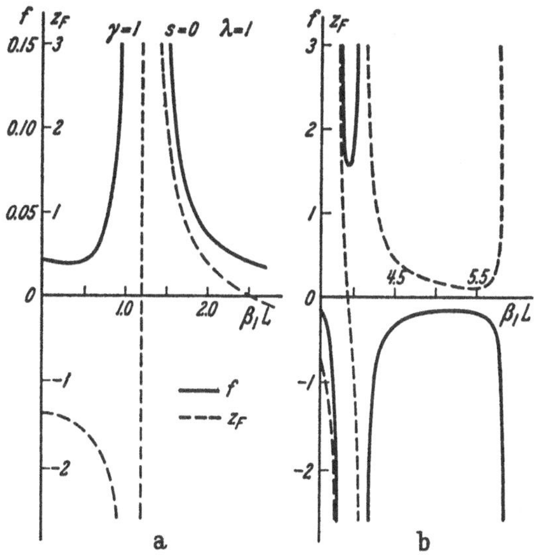

Fig. 28a and b. Region II. Focal length and z_F as functions of $\beta_1 L$ for $\gamma = 1$, $s = 0$, $\lambda = 1$. (a) Third branch; (b) fourth branch salient features of *Seeliger's* system [166]: in both systems, correction is applied at real line images. We shall not dwell further here on *Deltrap's* system which is described in detail by *Septier* [187]. Each member of the

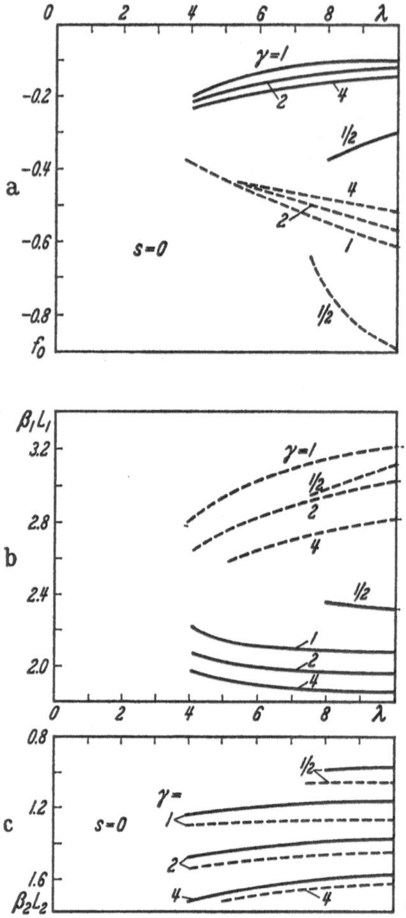

Fig. 29

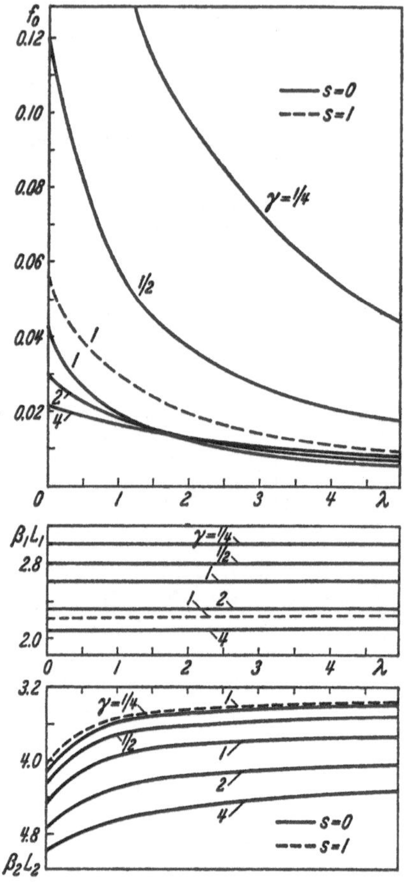

Fig. 30

Fig. 29a—c. Region C, $z_F = 0$. (a): the dependence of focal length upon γ and λ for $s = 0$; (b) and (c): the corresponding excitations (——: first branch; - - - -: second branch)

Fig. 30. Third branch. The dependence of focal length and excitations upon λ for $z_F = 0$ and various values of s and γ

quadruplet produced both quadrupole and octopole potentials, and the degree of correction achieved is illustrated in Fig. 33, together with a schematic view of the system.

Other systems. A number of other suggestions for quadrupole systems have been made, and we conclude this section with a brief review of these. Further details of those designed to combat spherical aberration are given in *Septier's* review article on "The struggle to overcome spherical aberration in electron optics" [187]. *Seeliger* and *Möllenstedt*

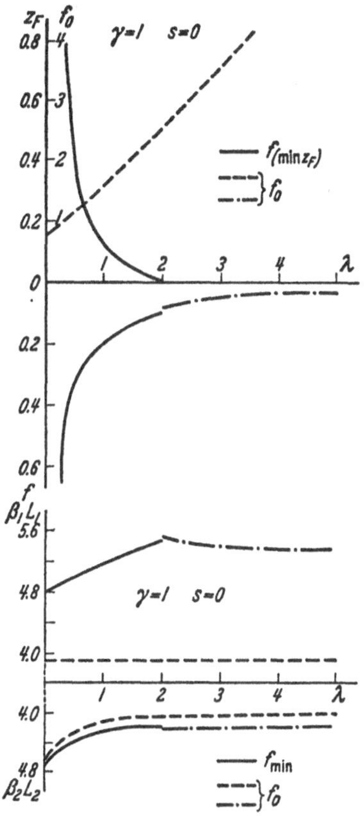

Fig. 31. Fourth branch. Focal length and excitations for $z_F = 0$ and the minimum focal lengths for $\gamma = 1$, $s = 0$

[166, 139, 140, 157] attempted to exploit *Scherzer's* suggestions of 1947 by building a system of cylindrical lenses and octopoles which was potentially capable of a higher resolution than a round lens system; the mechanical complexity of this system rendered it impractical, however. *Archard* [3, 4, 5] suggested that considerable simplification could be achieved by replacing the slit lenses by quadrupoles, and incorporating octopole elements into the quadrupoles.

A quadrupole triplet or quadruplet condenser has been described by *Recknagel* [149], who also mentions a triplet condenser investigated by *Schüler* [165]. More recently, *Bauer* [10] has designed and built a five-quadrupole electrostatic objective. The distance from object to image is 50 cm, with a magnification of 800 ×; the bore radius is 2 mm, the system is 8 cm long and the excitations lie between ±3 and ±7% of the accelerating potential. (In his latest account, the electrode voltages lie between ±4.8% and ±9% of the accelerating voltage.) The lens

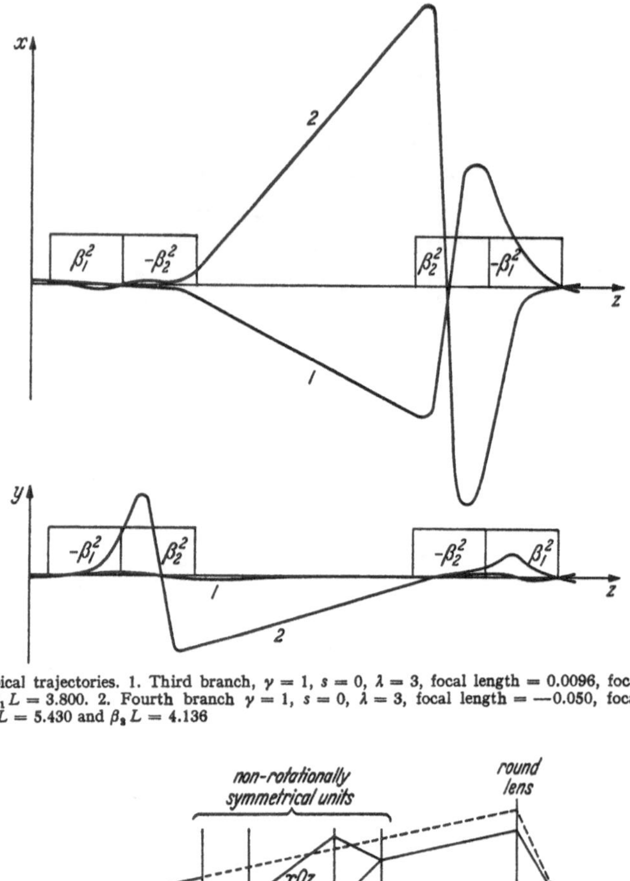

Fig. 32. Typical trajectories. 1. Third branch, $\gamma = 1$, $s = 0$, $\lambda = 3$, focal length = 0.0096, focal distance = 0.0490, $\beta_1 L = 3.800$. 2. Fourth branch $\gamma = 1$, $s = 0$, $\lambda = 3$, focal length = —0.050, focal distance = 0.001, $\beta_1 L = 5.430$ and $\beta_2 L = 4.136$

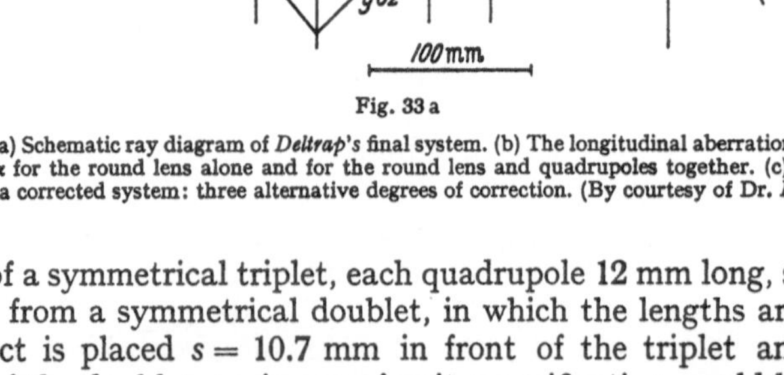

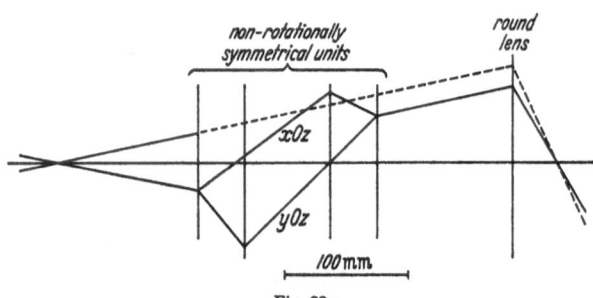

Fig. 33 a

Fig. 33a—c. (a) Schematic ray diagram of *Deltrap's* final system. (b) The longitudinal aberration as a function of the angle α for the round lens alone and for the round lens and quadrupoles together. (c) Longitudinal aberration in a corrected system: three alternative degrees of correction. (By courtesy of Dr. *Deltrap*)

consists of a symmetrical triplet, each quadrupole 12 mm long, separated by 4 mm from a symmetrical doublet, in which the lengths are 21 mm. The object is placed $s = 10.7$ mm in front of the triplet and in the absence of the doublet, an image of unit magnification would be formed at a distance s beyond the triplet. The high magnification is achieved by arranging that this image would fall near the immersion focus of the (stigmatic) doublet.

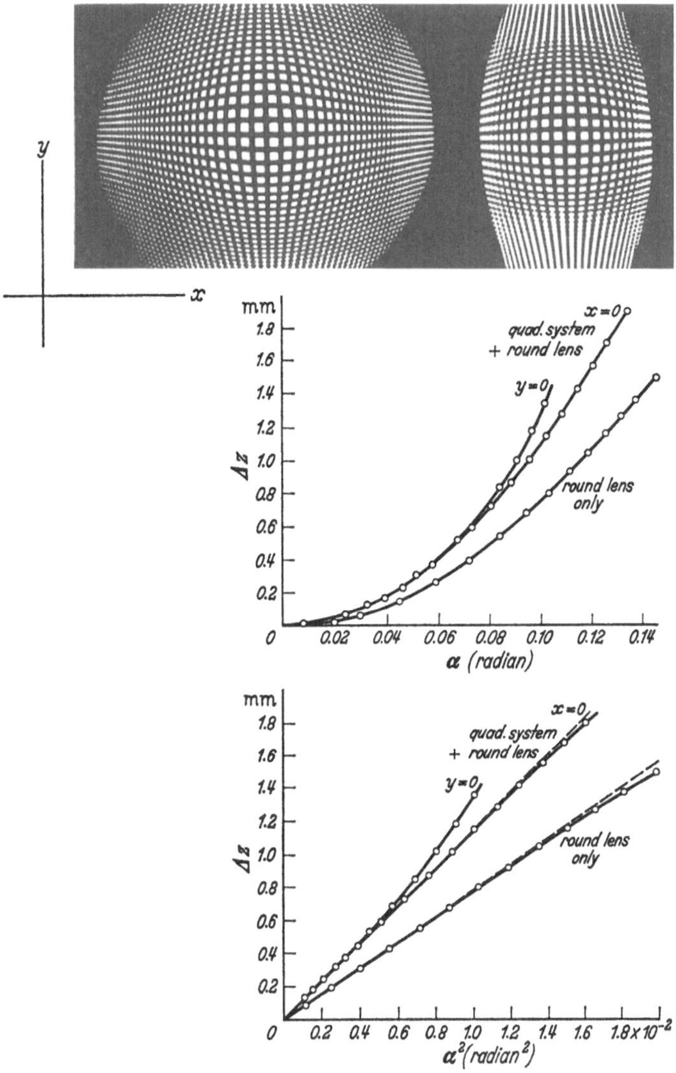

Fig. 33 b

All the quadrupoles discussed above do indeed consist of four electrodes or pole-pieces; quadrupole potential distributions may be produced by other means, though, and *Klemperer* has suggested that double-lipped cylinders might be useful in practice [116, 117; cf. 122a, 128a]; such cylinders are not terminated by a plane, but by a curve such that the departure of any point on the periphery is a function of $\cos 2\theta$ (cylindrical polar coordinates, see Fig. 34). A practical use for these "quadrupole

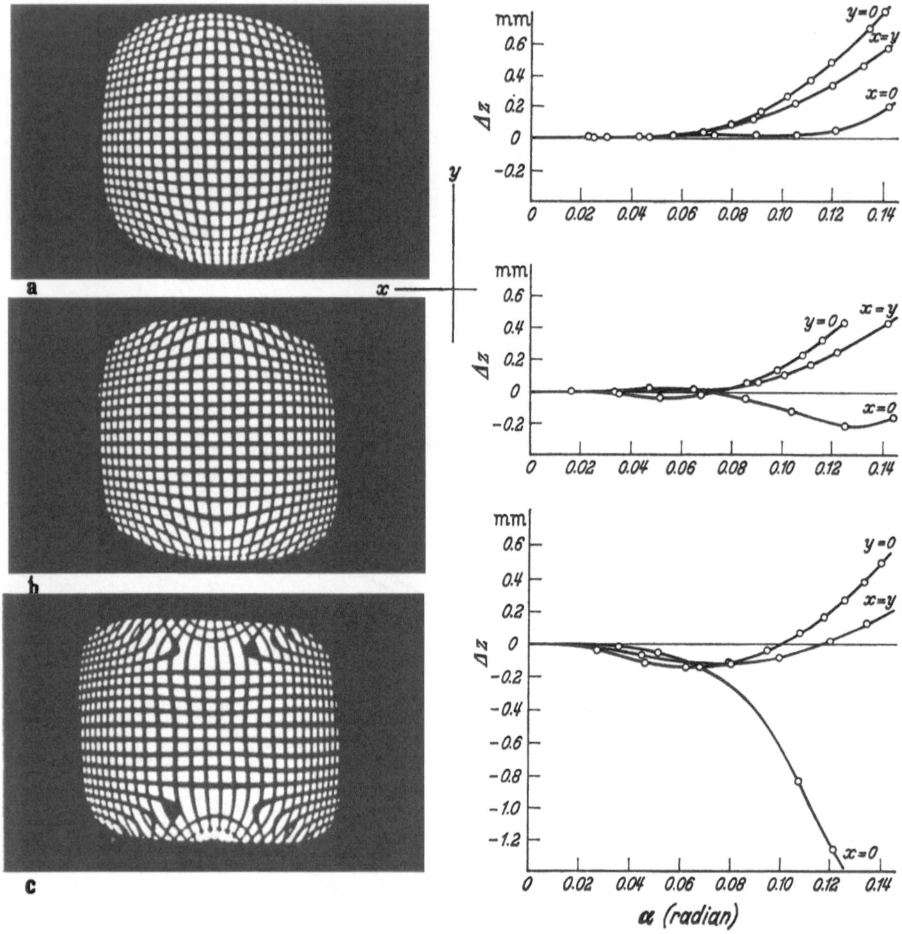

Fig. 33c

gaps" is described by *Boussard* and *Septier* [18b, 18c]. Finally, we should mention that one class of lenses producing a quadrupole component, cylindrical lenses, has been completely excluded from this survey; for a critical account of work on these lenses, see [94]. Octopoles are studied in most of the attempts to correct spherical aberration with their aid, and the associated caustics have been investigated by *Schleich* [161, 163]. *Strashkevich* has studied multipoles in general [203].

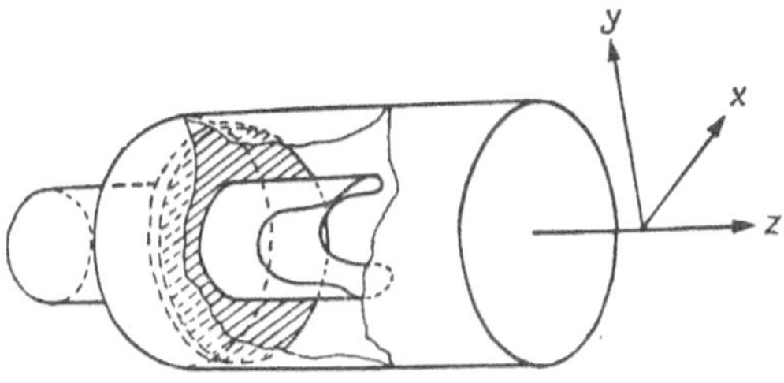

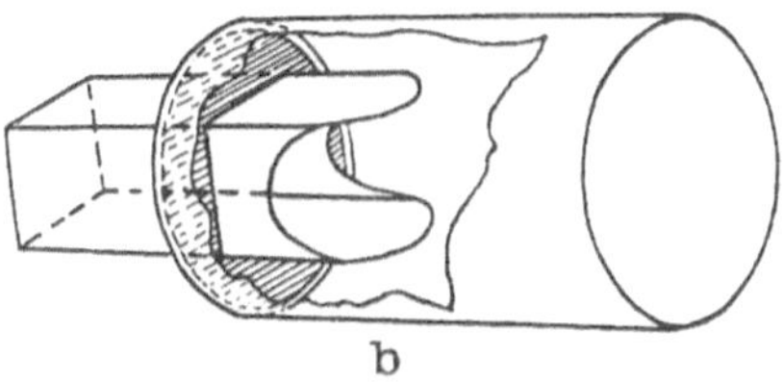

Fig. 34. *Klemperer's* double-lipped cylinders: An interlocking pair of such cylinders would produce a quadrupole potential distribution across the gap. (By courtesy of Prof. *Klemperer* and the Cambridge University Press)

5. Chromatic aberration

5.1 Introductory

This aberration is analysed by the same methods as all the other primary aberrations, discussed in § 3. It is accorded a separate section, however, because the chromatic aberration coefficients of a particular combination of magnetic and electrostatic quadrupoles can all vanish simultaneously, and lenses of this type have in consequence been much studied. That the chromatic aberration of a round lens can be corrected with the aid of quadrupoles and octopoles was discussed by *Scherzer* [156, 158], but it was in 1961 that *Kel'man* and *Yavor* [108] demonstrated that coincident electrostatic and magnetic quadrupoles (producing rectangular potentials of the same length) can be achromatic [110]. The solutions of the equations of motion

$$x'' + \beta^2 x = 0\,, \quad y'' - \beta^2 y = 0\,, \quad \beta^2 = -\frac{D_0 - 4\eta\, Q_0 \sqrt{\Phi}}{4\Phi}$$

are insensitive to changes in Φ if $\partial\beta^2/\partial\Phi$ vanishes, and hence, if $D_0 = 2\eta\, Q_0 \sqrt{\Phi}$; with $\beta_E^2 = D_0/4\Phi$ and $\beta_M^2 = \eta\, Q_0/\sqrt{\Phi}$, this condition may be written

$$D_0 = 2\eta\, Q_0 \sqrt{\Phi} \quad \text{or} \quad \beta_M^2 = 2\beta_E^2\,. \tag{5.1}$$

Septier later came to the same conclusion [186] and the full expressions for the chromatic aberration coefficients (*Hawkes* [91, 88]) show that the same result is true for any potential distributions, $Q(z)$, $D(z)$, provided that their z-dependence is similar, and that $D(z) = 2\eta\, Q(z)\sqrt{\Phi}$. The coefficients are in general given by

$$k_x\, x^{(c)} = t\, x_o + v\, x_a$$
$$k_y\, y^{(c)} = U\, y_o + W\, y_a \tag{5.2a}$$

in which

$$8t = h_x \int_a^c (\Phi^* - Q^*)\, g_x^2\, \mathrm{d}z - g_x \int_o^c (\Phi^* - Q^*)\, g_x h_x\, \mathrm{d}z$$

$$8v = h_x \int_a^c (\Phi^* - Q^*)\, g_x h_x\, \mathrm{d}z - g_x \int_o^c (\Phi^* - Q^*)\, h_x^2\, \mathrm{d}z$$

$$8U = h_y \int_a^c (\Phi^* + Q^*)\, g_y^2\, \mathrm{d}z - g_y \int_o^c (\Phi^* + Q^*)\, g_y h_y\, \mathrm{d}z$$

$$8W = h_y \int_a^c (\Phi^* + Q^*)\, g_y h_y\, \mathrm{d}z - g_y \int_o^c (\Phi^* + Q^*)\, h_y^2\, \mathrm{d}z \tag{5.2b}$$

with

$$\Phi^* = 3\Phi'^2/\Phi^{5/2}\,, \quad Q^* = 2(D - 2\eta\, Q\sqrt{\Phi})/\Phi^{3/2}\,. \tag{5.2c}$$

Thus when $\Phi =$ constant, $\Phi^* = 0$ and all the coefficients then vanish if $Q^* = 0$, $D = 2\eta\, Q\sqrt{\Phi}$. See also *Strashkevich* [213a].

Subsequent Russian work has been concerned with the aperture and other aberration coefficients of achromatic and mixed lenses, with the possibility of obtaining quadrupole lenses with negative chromatic aberration, and with the effect of departures from exact proportionally between $D(z)$ and $Q(z)$; in [109, 238, 240], the achromatism and the existence of negative values are demonstrated experimentally. Achromatic multipoles are analysed in [237].

5.2 Geometrical aberrations

The fullest study of the aberration coefficients is to be found in *Ovsyannikova* and *Yavor* [143], in which a number of earlier papers culminate [51, 56, 58]. As above, we set $\beta^2 = \beta_M^2 - \beta_E^2$, and with the functions $D(z)$, $Q(z)$ given by $D(z) = 4\Phi\,\beta_E^2 f(z)$, $\eta\, Q(z) = \sqrt{\Phi}\,\beta_M^2 f(z)$, the third-order equations take the form

$$x'' + \beta^2 f x = -\beta^2 \Big[\frac{1}{2} f x \{(n+3)\, x'^2 - (n-1)\, y'^2\} -$$
$$- (n+1) f y x' y' + \frac{1}{2} n f' x' (x^2 - y^2) - (n+1) f' x y y' +$$
$$+ \frac{1}{4} f'' x \Big\{\frac{n-1}{3} x^2 - (n+1) y^2\Big\} + \frac{1}{2} n (n-1) \beta^2 f^2 x (x^2 - y^2)\Big]$$

$$y'' - \beta^2 f y = \beta^2 \Big[-\frac{1}{2} f y \{(n-1) x'^2 - (n+3) y'^2\} -$$
$$- (n+1) f x x' y' - \frac{1}{2} n f' y' (x^2 - y^2) - (n+1) f' x y x' -$$
$$- \frac{1}{4} f'' y \Big\{ (n+1) x^2 - \frac{n-1}{3} y^2 \Big\} + \frac{1}{2} n (n-1) \beta^2 f^2 y (x^2 - y^2) \Big] . \tag{5.3}$$

The parameter n is a measure of the relative strengths of the magnetic and electrostatic lenses: $n = \beta_E^2/\beta^2 = \beta_E^2/(\beta_M^2 - \beta_E^2)$, so that for magnetic lenses alone, $n = 0$ and for electrostatic lenses alone, $n = -1$; for achromatic lenses, $n = 1$. The paraxial solutions s_x, s_y, T_x and T_y are employed, and variation of parameters yields the following expression for $x^{(3)}$ at the line image, z_i, where $T_x(z_i) = 0$:

$$x_i^{(3)} = \frac{1}{T'_{xi}} \{ (0030)^* x_o'^3 + (0012)^* x_o' y_o'^2 + (1020)^* x_o x_o'^2 +$$
$$+ (1002)^* x_o y_o'^2 + (0111)^* y_o x_o' y_o' + (2010)^* x_o^2 x_o' + (0210)^* y_o^2 x_o' +$$
$$+ (1101)^* x_o y_o y_o' + (3000)^* x_o^3 + (1200)^* x_o y_o^2 \} \tag{5.4}$$

in which

$$(0030)^* = \frac{1}{6} \int_o^i T_x'^4 \,\mathrm{d}z + \frac{1}{6} (3n^2 - 2n + 2) \beta^4 \int_o^i f^2 T_x^4 \,\mathrm{d}z$$

$$(0012)^* = -\frac{1}{2} T'^2_{xi} T_{yi} T'_{yi} + \frac{3}{2} \int_o^i T_x'^2 T_y'^2 \,\mathrm{d}z - \frac{1}{2} (n^2 - 2n -$$
$$- 2) \beta^4 \int_o^i f^2 T_x^2 T_y^2 \,\mathrm{d}z$$

$$(1020)^* = \frac{1}{4} (T'^2_{xi} - 1) + \frac{1}{2} \int_o^i T_x'^3 s_x' \,\mathrm{d}z + \frac{1}{2} (3n^2 - 2n +$$
$$+ 2) \beta^4 \int_o^i f^2 T_x^3 s_x \,\mathrm{d}z$$

$$(1002)^* = -\frac{1}{2} T'_{xi} s'_{xi} T_{yi} T'_{yi} - \frac{1}{4} (T'^2_{yi} - 1) + \frac{n+1}{4} \beta^2 f_i T'_{xi} s_{xi} T^2_{yi} +$$
$$+ \frac{3}{2} \int_o^i T'_x s'_x T_y'^2 \,\mathrm{d}z - \frac{1}{2} (n^2 - 2n - 2) \beta^4 \int_o^i f^2 T_x s_x T_y^2 \,\mathrm{d}z$$

$$(0111)^* = \frac{1}{2} (1 - T'^2_{xi}) - T'^2_{xi} T_{yi} s'_{yi} + 3 \int_o^i T_x'^2 T'_y s'_y \,\mathrm{d}z -$$
$$- (n^2 - 2n - 2) \beta^4 \int_o^i f^2 T_x^2 T_y s_y \,\mathrm{d}z$$

$$(2010)^* = \frac{1}{2} T_{xi}'^2 s_{xi} s_{xi}' + \frac{1}{2} \int_o^i T_x'^2 s_x'^2 \,dz + \frac{1}{2} (3n^2 - 2n + 2) \beta^4 \int_o^i f^2 T_x^2 s_x^2 \,dz$$

$$(0210)^* = -\frac{1}{2} T_{xi}'^2 s_{yi} s_{yi}' + \frac{3}{2} \int_o^i T_x'^2 s_y'^2 \,dz - \frac{1}{2} (n^2 - 2n - 2) \beta^4 \int_o^i f^2 T_x^2 s_y^2 \,dz$$

$$(1101)^* = -T_{xi}' s_{xi}' T_{yi} s_{yi}' + \frac{n+1}{2} \beta^2 f_i T_{xi}' s_{xi} T_{yi} s_{yi} - \beta^2 \int_o^i f (T_y s_y' - T_x s_x') \,dz + 3 \int_o^i T_x' s_x' T_y' s_y' \,dz - (n^2 - 2n - 2) \beta^4 \int_o^i f^2 T_x s_x T_y s_y \,dz$$

$$(3000)^* = \frac{1}{4} s_{xi}'^2 - \frac{n-1}{12} \beta^2 (f_i T_{xi}' s_{xi}^3 - f_o) + \frac{1}{6} \int_o^i T_x' s_x'^3 \,dz + \frac{1}{6} (3n^2 - 2n + 2) \beta^4 \int_o^i f^2 T_x s_x^3 \,dz$$

$$(1200)^* = -\frac{1}{4} s_{yi}'^2 - \frac{1}{2} T_{xi}' s_{xi}' s_{yi} s_{yi}' + \frac{n+1}{4} \beta^2 (f_i T_{xi}' s_{xi} s_{yi}^2 - f_o) + \frac{3}{2} \int_o^i T_x' s_x' s_y'^2 \,dz - \frac{1}{2} (n^2 - 2n - 2) \beta^4 \int_o^i f^2 T_x s_x s_y^2 \,dz . \tag{5.5}$$

To obtain the aberration coefficients in terms of x_a and y_a instead of x_o' and y_o', we substitute for the latter from*

$$x_a = x_o s_{xa} + x_o' T_{xa}, \quad y_a = y_o s_{ya} + y_o' T_{ya} \tag{5.6}$$

to give

$$x_i^{(3)} = \frac{1}{T_{xi}'} \{(0030)\, x_a^3 + (0012)\, x_a y_a^2 + (1020)\, x_o x_a^2 + (1002)\, x_o y_a^2 + (0111)\, y_o x_a y_a + (2010)\, x_o^2 x_a + (0210)\, y_o^2 x_a + (1101)\, x_o y_o y_a + (3000)\, x_o^3 + (1200)\, x_o y_o^2\} \tag{5.7}$$

---

* Thus $x(z) = s_x x_o + T_x x_o'$ becomes $x(z) = \left(s_x - \frac{s_{xa}}{T_{xa}} T_x\right) x_o + \frac{T_x}{T_{xa}} x_a$ which is identical with $g_x x_o + h_x x_a$; the aberration coefficients, $x^{(3)}$, are obtained correctly in the line-image plane where T_x and h_x vanish, but the formulae obtained by this procedure for $y_i^{(3)}$ would not be correct [93].

where now

$$(0030) = (0030)^*/T_{xa}^3$$
$$(0012) = (0012)^*/T_{xa}\, T_{ya}^2$$
$$(1020) = \{(1020)^* - 3(0030)^*\, s_{xa}/T_{xa}\}/T_{xa}^2$$
$$(1002) = \{(1002)^* - (0012)^*\, s_{xa}/T_{xa}\}/T_{ya}^2$$
$$(0111) = \{(0111)^* - 2(0012)^*\, s_{ya}/T_{ya}\}/T_{xa}\, T_{ya}$$
$$(2010) = \{3(0030)^*\, s_{xa}^2/T_{xa}^2 - 2(1020)^*\, s_{xa}/T_{xa} + (2010)^*\}/T_{xa}$$
$$(0210) = \{(0012)^*\, s_{ya}^2/T_{ya}^2 - (0111)^*\, s_{ya}/T_{ya} + (0210)^*\}/T_{xa}$$
$$(1101) = \{2(0012)^*\, s_{xa}\, s_{ya}/T_{xa}\, T_{ya} - 2(1002)^*\, s_{ya}/T_{ya} - - (0111)^*\, s_{xa}/T_{xa} + (1101)^*\}/T_{ya}$$
$$(3000) = - (0030)^*\, s_{xa}^3/T_{xa}^3 + (1020)^*\, s_{xa}^2/T_{xa}^2 - (2010)^*\, s_{xa}/T_{xa} + (3000)^*$$
$$(1200) = - (0012)^*\, s_{xa}\, s_{ya}^2/T_{xa}\, T_{ya}^2 + (1002)^*\, s_{ya}^2/T_{ya}^2 + + (0111)^*\, s_{xa}\, s_{ya}/T_{xa}\, T_{ya} - (0210)^*\, s_{xa}/T_{xa} - - (1101)^*\, s_{ya}/T_{ya} + (1200)^* \,. \tag{5.8}$$

In an arbitrary plane, the aperture aberrations are given by $x^{(3)}(z) = (30)^*\, x_o'^3 + (12)^*\, x_o'\, y_o'^2$; $y^{(3)}(z) = (03)^*\, y_o'^3 + (21)^* x_o'^2\, y_o'$, where [59]

$$(30)^* = T_x \left\{\frac{1}{4}(T_x'^2 - 1) - \frac{n-1}{12}\beta^2 f\, T_x^2 - \frac{1}{2}\beta^2 \int_o^z f\, s_x\, T_x\, T_x'^2\, dz - - \frac{1}{6}\beta^4 (3n^2 - 2n + 2) \int_o^z f^2\, s_x\, T_x^3\, dz\right\} + + s_x \left\{\frac{1}{2}\beta^2 \int_o^z f\, T_x^2\, T_x'^2\, dz + \frac{1}{6}\beta^4 (3n^2 - 2n + 2) \int_o^z f^2\, T_x^4\, dz\right\}$$

$$(12)^* = T_x \left\{-\frac{1}{4}(T_y'^2 + 3) - \frac{1}{2} T_y\, T_x'\, T_y'\Big/ T_x + \frac{n+1}{4}\beta^2 f\, T_y^2 - - \frac{3}{2}\int_o^z s_x'\, T_x'\, T_y'^2\, dz + \frac{1}{2}\beta^4 (n^2 - 2n - 2) \int_o^z f^2\, s_x\, T_x\, T_y^2\, dz\right\} + + s_x \left\{\frac{3}{2}\int_o^z T_x'^2\, T_y'^2\, dz - \frac{1}{2}\beta^4 (n^2 - 2n - 2) \int_o^z f^2\, T_x^2\, T_y^2\, dz\right\}$$

$$(03)^* = T_y \left\{\frac{1}{4}(T_y'^2 - 1) - \frac{n-1}{12}\beta^2 f\, T_y^2 + \frac{1}{2}\beta^2 \int_o^z f\, s_y\, T_y\, T_y'^2\, dz - - \frac{1}{6}\beta^4 (3n^2 - 2n + 2) \int_o^z f^2\, s_y\, T_y^3\, dz\right\} + + s_y \left\{-\frac{1}{2}\beta^2 \int_o^z f\, T_y^2\, T_y'^2\, dz + \frac{1}{6}\beta^4 (3n^2 - 2n + 2) \int_o^z f^2\, T_y^4\, dz\right\}$$

$$(21)^* = T_y \left\{ -\frac{1}{4}(T_x'^2 + 3) - \frac{1}{2} T_x T_x' T_y' \Big/ T_y - \frac{n+1}{4} \beta^2 f T_x^2 - \right.$$

$$- \frac{3}{2} \int_o^z s_y' T_x'^2 T_y' \, dz + \frac{1}{2} \beta^4 (n^2 - 2n - 2) \left. \int_o^z f^2 s_y T_x^2 T_y \, dz \right\} +$$

$$+ s_y \left\{ \frac{3}{2} \int_o^z T_x'^2 T_y'^2 \, dz - \frac{1}{2} \beta^4 (n^2 - 2n - 2) \int_o^z f^2 T_x^2 T_y^2 \, dz \right\}. \tag{5.9}$$

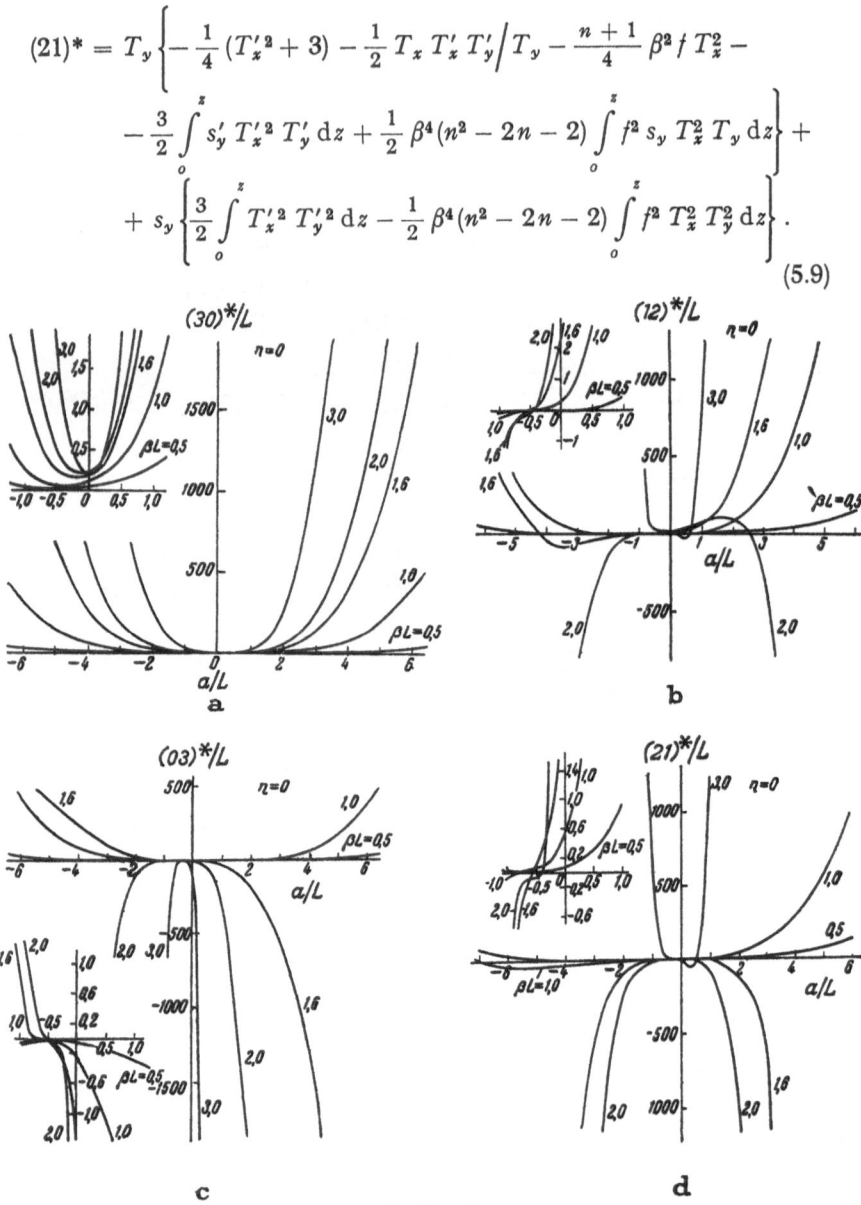

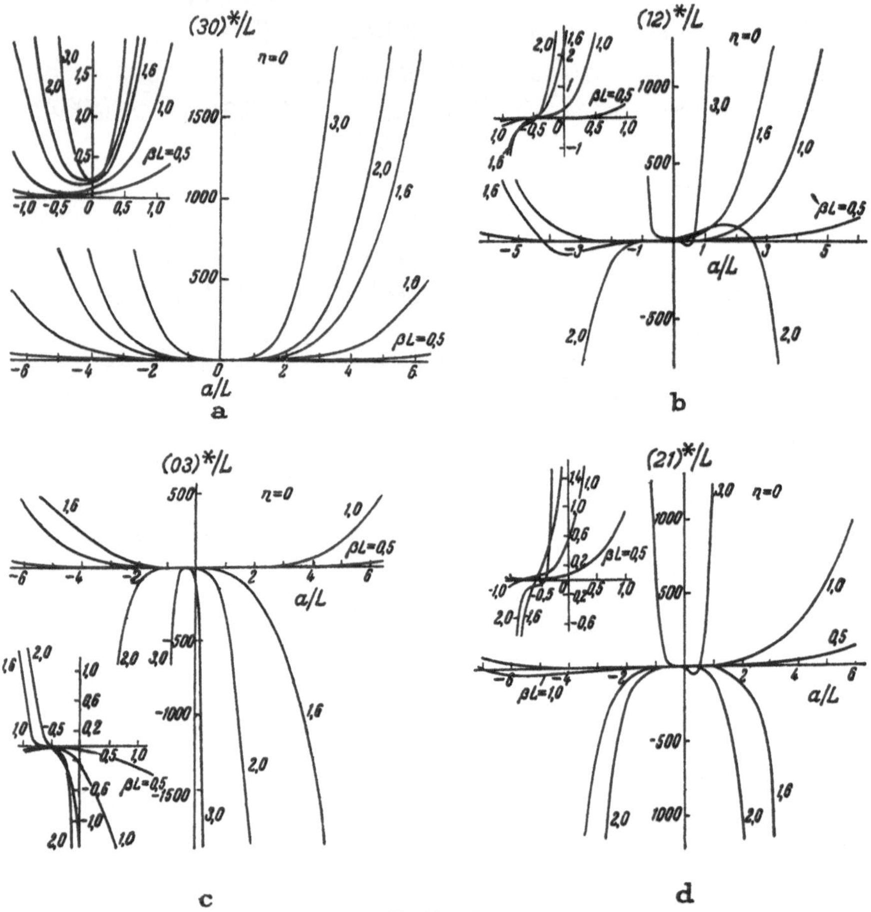

Fig. 35a—d

Fig. 35a—h. The dependence of the aperture aberration coefficients upon the position of the object, using the rectangular model; (a) — (d), magnetic quadrupole; (e) — (h), electrostatic quadrupole. The distance of the object plane from the beginning of the rectangle is *a*. (After [60c] with grateful acknowledgment; read (03) for (30) in (g)).

Formulae for the four aperture aberration coefficients applicable in any plane are also given in *Hawkes* [90]. [The parameter p in [90] is related to n by the formula $p = (n+1)/n$.] Expressions in the above form [no derivatives of $f(z)$ under the integrals] can be obtained straightforwardly from equations (3.43)*. In the remainder of [143], the case $f(z) = 1$ and the short lens are examined, and the dependence of the

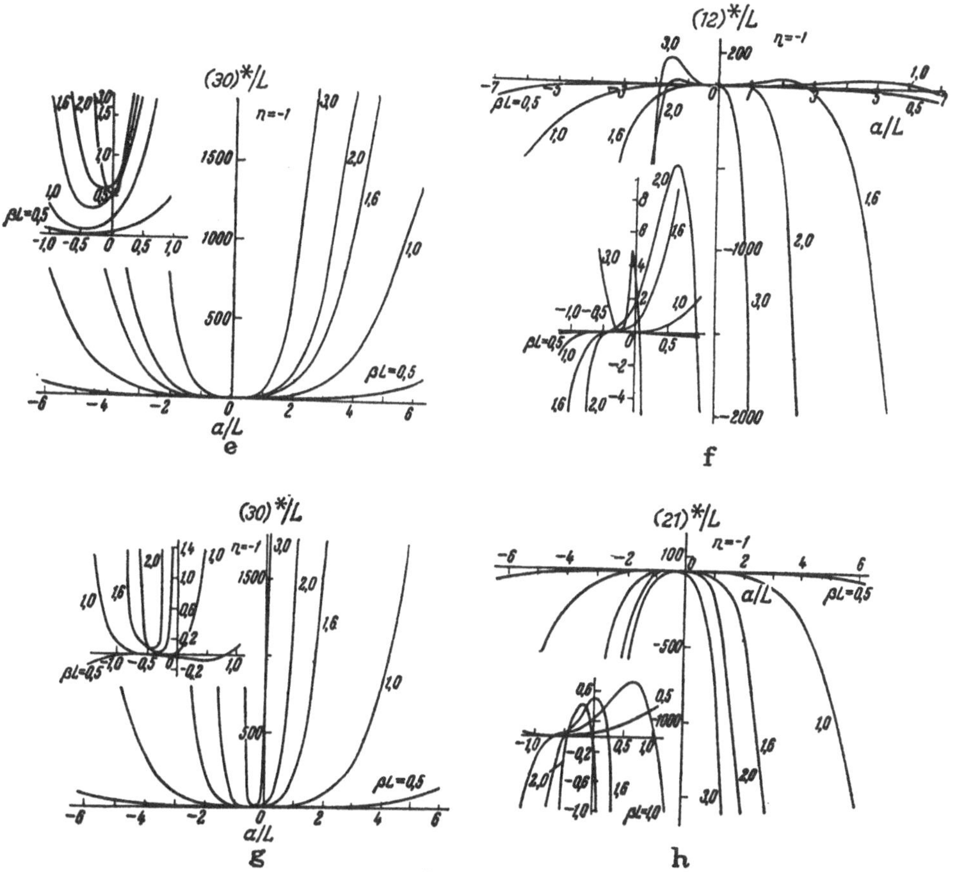

Fig. 35 e—h

coefficients upon the position of the aperture is briefly considered. In [60d], formulae for the aperture aberration coefficients are listed, in which the object positions in the $x - z$ and $y - z$ planes differ; the field is assumed to be of finite extent. Fifth order aperture aberrations are mentioned in [64a]. Graphs showing how the coefficients depend upon object position are plotted in [60c], see Fig. 35.

The forms of the aperture aberration coefficients using the rectangular model and the bell-shaped model are set out in [56, 60] and [59, 60a] respectively.

(i) *The rectangular model.* With $x_i^{(3)} = \{(30)^* x_o'^3 + (12)^* x_o' y_o'^2\}/T'_{xi}$ and $y_i^{(3)} = \{(03)^* y_o'^3 + (21)^* x_o'^2 y_o'\}/T'_{yi}$ where $z = z_i$ is the (real) line

* Put $\varrho_1 = -8$ in (3.34) to obtain *Dymnikov's* formulae.

image at which $T_x(z_i) = 0$ and $z = z_j$ is the (virtual line) image at which the tangent to T_y in image space crosses the axis, *Dymnikov* et al. find

$$16\,\frac{(30)^*}{L} = (1+\beta^2u^2)^2 - (1-6\beta^2u^2+\beta^4u^4)\,\frac{\sin 4\beta L}{4\beta L} +$$
$$+\frac{u}{L}(1-\beta^2u^2)(1-\cos 4\beta L) + \frac{1}{3}(2-2n+3n^2)\Big\{3(1+\beta^2u^2)^2 -$$
$$-4(1-\beta^4u^4)\,\frac{\sin 2\beta L}{2\beta L} + (1-6\beta^2u^2+\beta^4u^4)\,\frac{\sin 4\beta L}{4\beta L} +$$
$$+4\frac{u}{L}(1+\beta^2u^2)(1-\cos 2\beta L) - \frac{u}{L}(1-\beta^2u^2)(1-\cos 4\beta L)\Big\}$$

$$16\,\frac{(12)^*}{L} = 6(1-\beta^4u^4) + 2\frac{u}{L}\{3-(1-\beta^2u^2)\cos 2\beta L +$$
$$+(1+\beta^2u^2)\,\mathrm{Ch}\,2\beta L - (2-3\beta^2u^2)\sin 2\beta L\,\mathrm{Sh}\,2\beta L +$$
$$-(3+2\beta^2u^2)\cos 2\beta L\,\mathrm{Ch}\,2\beta L\} - \{2(1-\beta^2u^2)^2 +$$
$$+(5-4\beta^2u^2-5\beta u^4)\,\mathrm{Ch}\,2\beta L\}\,\frac{\sin 2\beta L}{2\beta L} +$$
$$+\{2(1+\beta^2u^2)^2 - (1+20\beta^2u^2-\beta^4u^4)\cos 2\beta L\}\,\frac{\mathrm{Sh}\,2\beta L}{2\beta L} + N$$

$$16\,\frac{(03)^*}{L} = (1-\beta^2u^2)^2 - (1+6\beta^2u^2+\beta^4u^4)\,\frac{\mathrm{Sh}\,4\beta L}{4\beta L} +$$
$$+\frac{u}{L}(1+\beta^2u^2)(1-\mathrm{Ch}\,4\beta L) + \frac{1}{3}(2-2n+3n^2)\Big\{3(1-\beta^2u^2)^2 -$$
$$-4(1-\beta^4u^4)\,\frac{\mathrm{Sh}\,2\beta L}{2\beta L} + (1+6\beta^2u^2+\beta^4u^4)\,\frac{\mathrm{Sh}\,4\beta L}{4\beta L} +$$
$$+4\frac{u}{L}(1-\beta^2u^2)(1-\mathrm{Ch}\,2\beta L) - \frac{u}{L}(1+\beta^2u^2)(1-\mathrm{Ch}\,4\beta L)\Big\}$$

$$16\,\frac{(21)^*}{L} = 6(1-\beta^4u^4) + 2\frac{u}{L}\{3+(1-\beta^2u^2)\cos 2\beta L -$$
$$-(1+\beta^2u^2)\,\mathrm{Ch}\,2\beta L + (2+3\beta^2u^2)\sin 2\beta L\,\mathrm{Sh}\,2\beta L -$$
$$-(3-2\beta^2u^2)\cos 2\beta L\,\mathrm{Ch}\,2\beta L\} + \{2(1-\beta^2u^2)^2 -$$
$$-(1-20\beta^2u^2-\beta^4u^4)\,\mathrm{Ch}\,2\beta L\}\,\frac{\sin 2\beta L}{2\beta L} - \{2(1+\beta^2u^2) +$$
$$+(5+4\beta^2u^2-5\beta^4u^4)\cos 2\beta L\}\,\frac{\mathrm{Sh}\,2\beta L}{2\beta L} + N \qquad (5.10)$$

in which

$$N = (2+2n-n^2)\Big[-2(1-\beta^4u^4) - 2\frac{u}{L}\{1-(1-\beta^2u^2)\cos 2\beta L -$$
$$-(1+\beta^2u^2)\,\mathrm{Ch}\,2\beta L - \beta^2u^2\sin 2\beta L\,\mathrm{Sh}\,2\beta L +$$
$$+\cos 2\beta L\,\mathrm{Ch}\,2\beta L\} + \{2(1-\beta^2u^2)^2 -$$
$$-(1-4\beta^2u^2-\beta^4u^4)\,\mathrm{Ch}\,2\beta L\}\,\frac{\sin 2\beta L}{2\beta L} +$$
$$+\{2(1+\beta^2u^2)^2 - (1+4\beta^2u^2-\beta^4u^4)\cos 2\beta L\}\,\frac{\mathrm{Sh}\,2\beta L}{2\beta L}\Big].$$

L denotes the effective length and u is the object distance measured from the beginning of the rectangle. For high magnifications, $\beta u \to \cot\beta L$ which leads to considerable simplification, and for very large object distance (parallel incident rays)*

$$x_i^{(3)} = \frac{x_o}{L}\left\{(\overline{30})\left(\frac{x_o}{L}\right)^2 + (\overline{12})\left(\frac{y_o}{L}\right)^2\right\}$$

$$y_j^{(3)} = \frac{y_o}{L}\left\{(\overline{03})\left(\frac{y_o}{L}\right)^2 + (\overline{21})\left(\frac{x_o}{L}\right)^2\right\}$$

in which $(\overline{\alpha\,\beta})$ are obtained from $(\alpha\,\beta)$ in the limit $u \to \infty$, $x_o'\,u = x_o$, $y_o'\,u = y_o$. The papers conclude with a comparison with the results of *Septier* and *van Acker* [180] for electrostatic quadrupoles; theory and experiment agree, for the case compared, quite well.

(ii) *The bell-shaped model.* Again calculating the aberrations in the real line-image plane $z = z_i$ and the virtual line-image plane $z = z_j$, we find that for an object point on the axis, the aberrations in $z = z_j$, defined by

$$y_j^{(3)} = y_i^{(3)} - \frac{y'_{\mathrm{asymp}}}{y_{\mathrm{asymp}}}\, y_i'^{(3)} \tag{5.11a}$$

are given by

$$y_j^{(3)} = -\frac{1}{T'_{yi}}\int_o^i T_y(z)\, Y(z)\,\mathrm{d}z \tag{5.11b}$$

in which $Y(z)$ denotes the third-order terms in the equation of motion into which the paraxial solutions have been substituted. With $f(z) = \{1 + (z-\zeta)^2/d^2\}^{-2}$, the coefficients in $x_i^{(3)} = \{(30)^*\, x_o'^3 + (12)^*\, x_o'\, y_o'^2\}/T'_{xi}$, $y_j^{(3)} = \{(03)^*\, y_o'^3 + (21)^*\, x_o'^2\, y_o'\}/T'_{yi}$ are found to be as follows; we have written $\omega_x^2 = 1 + \beta^2 d^2$ and $\omega_y^2 = 1 - \beta^2 d^2$, $z - \zeta = d\cot\varphi$. (The signs in ω_x^2 and ω_y^2 seem to have become reversed because we follow the Russian convention whereby β^2 denotes $\beta_M^2 - \beta_E^2$.)

$$\frac{(30)^*}{d} = \frac{1}{32\sin^4\varphi_o}\left[(\omega_x^2 - 1)(\omega_x^2 + 3)\frac{k\pi}{\omega_x^5} + 2\,\frac{7-\omega_x^2}{4\omega_x^2 - 1}(\sin 2\varphi_o - \sin 2\varphi_i) + \right.$$
$$\left. + (2 - 2n + 3n^2)(\omega_x^2 - 1)\left\{(\omega_x^2 - 1)\frac{k\pi}{\omega_x^5} - \frac{2}{4\omega_x^2 - 1}(\sin 2\varphi_o - \sin 2\varphi_i)\right\}\right]$$

$$\frac{(12)^*}{d} = \frac{1}{32\,\omega_y^2\sin^4\varphi_o}\left[-4\left\{\left(1 - \cos 2k\pi\frac{\omega_y}{\omega_x}\right)\sin 2\varphi_i + \right.\right.$$
$$\left. + \omega_y(1 - \cos 2\varphi_i)\sin 2k\pi\frac{\omega_y}{\omega_x}\right\} + 3(\omega_x^2 - 1)\left[-2(\omega_x^2 - 1)\frac{k\pi}{\omega_x^3} + \right.$$
$$+ \frac{1}{\omega_x^2 - 1}\left\{-\frac{4\omega_y^2(\omega_x^2\omega_y^2 + 2)}{4\omega_x^2\omega_y^2 - 1}\sin 2\varphi_o + (\omega_y^2 + 3)\sin 2\varphi_i + \right.$$
$$\left. + \frac{3\omega_y^2 + 1}{2\omega_y}\sin 2k\pi\frac{\omega_y}{\omega_x}\right\} - \frac{3}{4\omega_x^2\omega_y^2 - 1}\left\{(4\omega_y^2 - 1)\sin 2\varphi_i \times \right.$$

* Explicit formulae for the aperture aberrations in both cases are given in [56]; as expected, the expressions for (30)* differ by a factor f_x^3.

$$\times \cos 2k\pi \frac{\omega_y}{\omega_x} + \omega_y (2\omega_y^2 + 1) \cos 2\varphi_i \sin 2k\pi \frac{\omega_y}{\omega_x}\Big\}\Big] +$$

$$+ (2 + 2n - n^2)(\omega_x^2 - 1)\Big[2(\omega_x^2 - 1)\frac{k\pi}{\omega_x^3} - \frac{4\omega_y^2(\omega_x^2 - 1)}{4\omega_x^2\omega_y^2 - 1}\sin 2\varphi_o +$$

$$+ \sin 2\varphi_i - \frac{1}{2\omega_y}\sin 2k\pi\frac{\omega_y}{\omega_x} - \frac{1}{4\omega_x^2\omega_y^2 - 1}\Big\{(4\omega_y^2 - 1)\sin 2\varphi_i \times$$

$$\times \cos 2k\pi\frac{\omega_y}{\omega_x} + \omega_y(2\omega_y^2 + 1)\cos 2\varphi_i \sin 2k\pi\frac{\omega_y}{\omega_x}\Big\}\Big]\Big]$$

$$\frac{(03)^*}{d} = \frac{\omega_x^2 - 1}{32\,\omega_y^4 \sin^4\varphi_o}\Big[-\frac{1}{3}\Big\{2(2 + n)\,\omega_y(1 - \cos 2\varphi_i) \times$$

$$\times \left(1 - \cos 2k\pi\frac{\omega_y}{\omega_x}\right)\sin 2k\pi\frac{\omega_y}{\omega_x} +$$

$$+ (4 - n)\left(\cos 4k\pi\frac{\omega_y}{\omega_x} - 4\cos 2k\pi\frac{\omega_y}{\omega_x} + 3\right)\sin 2\varphi_i\Big\} -$$

$$- (\omega_y^2 + 3)\frac{k\pi}{\omega_x} + \frac{2\omega_y^4(5 + \omega_x^2)}{(\omega_x^2 - 1)(4\omega_y^2 - 1)}\sin 2\varphi_o + \frac{1 + \omega_x^2}{2}\sin 2\varphi_i +$$

$$+ \frac{2}{\omega_y}\sin 2k\pi\frac{\omega_y}{\omega_x} - \frac{\omega_x^2 - 1}{4\omega_y}\sin 4k\pi\frac{\omega_y}{\omega_x} -$$

$$- \frac{2}{\omega_x^2 - 1}\Big\{(\omega_y^2 + 1)\sin 2\varphi_i \cos 2k\pi\frac{\omega_y}{\omega_x} + 2\omega_y \cos 2\varphi_i \sin 2k\pi\frac{\omega_y}{\omega_x}\Big\} -$$

$$- \frac{1}{2(4\omega_y^2 - 1)}\Big\{(5\omega_y^2 + 1)\sin 2\varphi_i \cos 4k\pi\frac{\omega_y}{\omega_x} +$$

$$+ 2\omega_y(\omega_y^2 + 2)\cos 2\varphi_i \sin 4k\pi\frac{\omega_y}{\omega_x}\Big\} +$$

$$+ \frac{1}{3}(2 - 2n + 3n^2)(\omega_x^2 - 1)\Big\{\frac{3k\pi}{\omega_x} + \frac{6\omega_y^4}{(\omega_x^2 - 1)(4\omega_y^2 - 1)}\sin 2\varphi_o +$$

$$+ \frac{3}{2}\sin 2\varphi_i - \frac{2}{\omega_y}\sin 2k\pi\frac{\omega_y}{\omega_x} + \frac{1}{4\omega_y}\sin 4k\pi\frac{\omega_y}{\omega_x} -$$

$$- \frac{2}{\omega_x^2 - 1}\left(\sin 2\varphi_i \cos 2k\pi\frac{\omega_y}{\omega_x} + \omega_y \cos 2\varphi_i \sin 2k\pi\frac{\omega_y}{\omega_x}\right) -$$

$$- \frac{1}{2(4\omega_y^2 - 1)}\left(\sin 2\varphi_i \cos 4k\pi\frac{\omega_y}{\omega_x} + 2\omega_y \cos 2\varphi_i \sin 4k\pi\frac{\omega_y}{\omega_x}\right)\Big\}\Big]$$

$$\frac{(21)^*}{d} = \frac{(12)^*}{d} - \frac{1}{8\omega_y^2\sin^4\varphi_o}\Big\{\left(1 - \cos 2k\pi\frac{\omega_y}{\omega_x}\right)\sin 2\varphi_i +$$

$$+ \omega_y(1 - \cos 2\varphi_i)\sin 2k\pi\frac{\omega_y}{\omega_x}\Big\}. \tag{5.12}$$

We recall that k is an integer, $\varphi_o - \varphi_i = k\pi/\omega_x$.

For very high magnification, $\varphi_i \to 0$ and $\varphi_o = k\pi/\omega_x$, so that

$$\frac{(30)^*}{d} = \frac{1}{32\sin^4\dfrac{k\pi}{\omega_x}}\Big[(\omega_x^2 - 1)(\omega_x^2 + 3)\frac{k\pi}{\omega_x^5} + \frac{2(7 - \omega_x^2)}{4\omega_x^2 - 1}\sin\frac{2k\pi}{\omega_x} +$$

$$+ (2 - 2n + 3n^2)(\omega_x^2 - 1)\Big\{(\omega_x^2 - 1)\frac{k\pi}{\omega_x^5} - \frac{2}{4\omega_x^2 - 1}\sin\frac{2k\pi}{\omega_x}\Big\}\Big]$$

$$\frac{(03)^*}{d} = \frac{1}{32\sin^4\dfrac{k\pi}{\omega_x}}\,\frac{\omega_x^2-1}{\omega_y^4}\Bigg[-(\omega_y^2+3)\frac{k\pi}{\omega_x} + \frac{2\omega_y^2(\omega_x^2+5)}{(\omega_x^2-1)(4\omega_y^2-1)}\sin\frac{2k\pi}{\omega_x} -$$
$$-\frac{2(3\omega_y^2-1)}{\omega_y(\omega_x^2-1)}\sin 2k\pi\frac{\omega_y}{\omega_x} - \frac{13\omega_y^2-1}{4\omega_y(4\omega_y^2-1)}\sin 4k\pi\frac{\omega_y}{\omega_x} +$$
$$+\frac{2-2n+3n^2}{3}(\omega_y^2-1)\left\{\frac{3k\pi}{\omega_x} - \frac{2}{\omega_y(\omega_x^2-1)}\sin 2k\pi\frac{\omega_y}{\omega_x} +\right.$$
$$\left.+\frac{6\omega_y^4}{(\omega_x^2-1)(4\omega_y^2-1)}\sin\frac{2k\pi}{\omega_x} - \frac{1}{4\omega_y(4\omega_y^2-1)}\sin 4k\pi\frac{\omega_y}{\omega_x}\right\}\Bigg]$$

$$\frac{(21)^*}{d} = \frac{(12)^*}{d} = \frac{1}{32\sin^4\dfrac{k\pi}{\omega_x}}\,\frac{\omega_x^2-1}{\omega_y^2}\Bigg[-6(\omega_x^2-1)\frac{k\pi}{\omega_x^3} -$$
$$-\frac{3}{(\omega_x^2-1)(4\omega_x^2\omega_y^2-1)}\left\{4\omega_y^2(\omega_x^2\omega_y^2+2)\sin\frac{2k\pi}{\omega_x} -\right.$$
$$\left.-\frac{14\omega_y^4-\omega_y^2-1}{2\omega_y}\sin 2k\pi\frac{\omega_y}{\omega_x}\right\} + (2+2n-n^2)(\omega_x^2-1)\left\{\frac{2k\pi}{\omega_x^3} -\right.$$
$$\left.-\frac{1}{4\omega_x^2\omega_y^2-1}\left(4\omega_y^2\sin\frac{2k\pi}{\omega_x} + \frac{10\omega_y^2-1}{2\omega_y(\omega_x^2-1)}\sin 2k\pi\frac{\omega_y}{\omega_x}\right)\right\}\Bigg]. \quad (5.13)$$

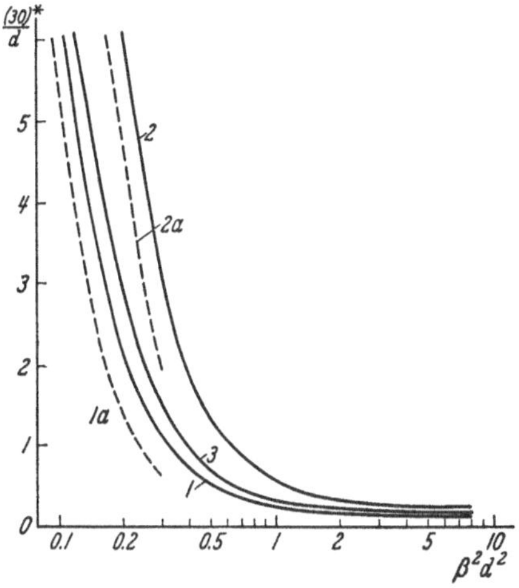

Fig. 36. The three aperture aberration coefficients as functions of $\beta^2 d^2$ for the first line image. 1. Magnetic quadrupole; 2: electrostatic quadrupole; 3: achromatic quadrupole; 4: combined lens for which $n = -2$ or $+4$. The dotted lines show the values obtained with the thin lens approximation. (After [59], with grateful acknowledgement)

The curves showing $(30)^*/d$, $(21)^*/d$ and $(03)^*/d$ as functions of $\beta^2 d^2$ for the first line image ($k = 1$) and for $n = 0$ (electrostatic quadrupoles), $n = -1$ (magnetic) and $n = 1$ (achromatic) are reproduced in Fig. 36. In the expressions for these coefficients, n appears as $(2 - 2n + 3n^2)$ in $(30)^*$ and $(03)^*$ and as $(2 + 2n - n^2)$ in $(12)^*$; the curves showing the coefficients as functions of n for two values of $\beta^2 d^2$ are thus symmetrical about $n = 1/3$ and $n = 1$, respectively. Fig. 37 illustrates the aberration coefficients as functions of n.

Consider now two bell-shaped distributions, distance D apart, with equal values of d. If they are arranged so that the virtual image of an object placed at the x-focus of the first lens lies close to the y-focus of

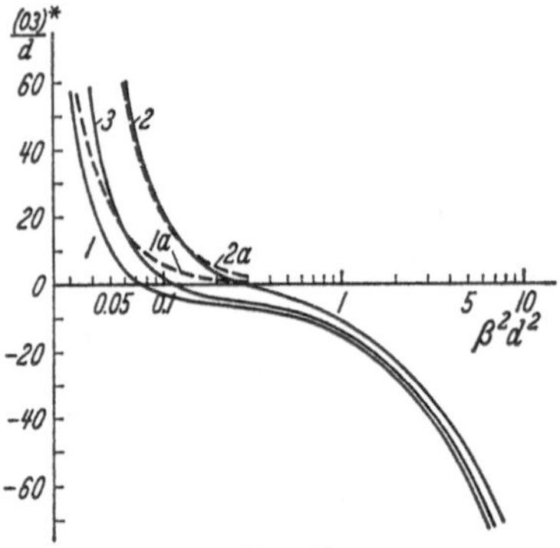

Fig. 36 b

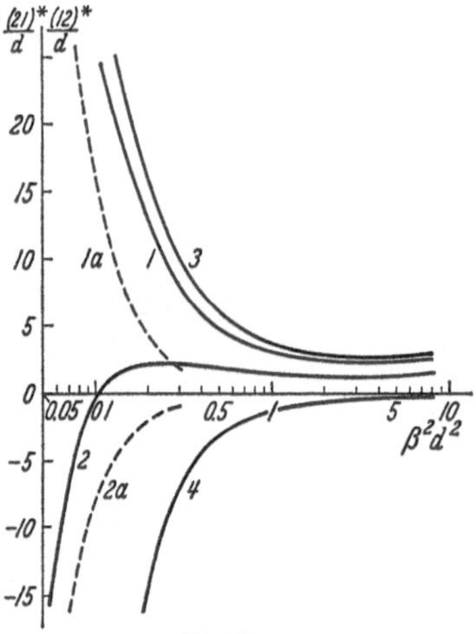

Fig. 36 c

the second, we have $\cot(\pi/\omega_{y2}) = \omega_{y1} \cot(\pi\, \omega_{y1}/\omega_{x1}) - D/d$. The aggregate aperture aberrations, $(\overline{\alpha\, \beta})$, are given by $(\overline{30}) = (30)_1^* +$

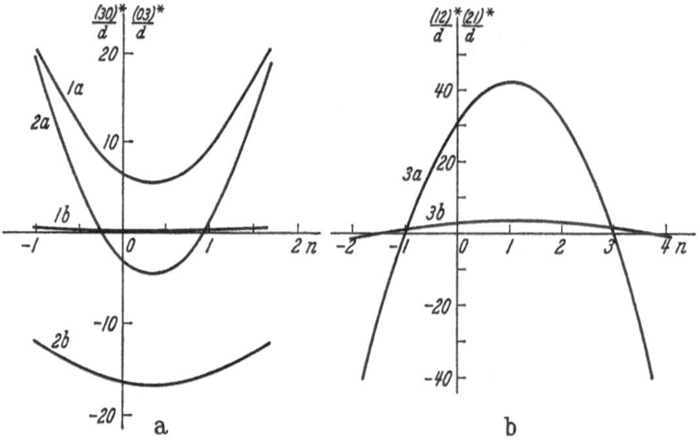

Fig. 37 a and b. The dependence of the aberration coeffcients upon n. 1: (30)*; 2: (03)*; 3: (21)*; for curves (a) $\beta^2 d^2 = 0.10$ and for curves (b), $\beta^2 d^2 = 1.10$. (After [59])

$+ (03)_2^* T'^4_{xi}$; $(\overline{03}) = (03)_1^* + (30)_2^* T'^4_{yi}$; $(\overline{12}) = (12)_1^* + (21)_2^* T'^2_{xi} T'^2_{yi}$; $(\overline{21}) = (21)_1^* + (12)_2^* T'^2_{xi} T'^2_{yi}$.

Dymnikov, Fishkova and *Yavor* find excitations and separations such that $(\overline{03})$ and $(\overline{21})$ vanish simultaneously; for $D/d = 2$ and $n_2 = 0$, they find $n_1 = -1.75$, $\beta_1^2 d^2 = 2.86$ and $\beta_2^2 d^2 = 1.21$, while for $D/d = 3$, $n_2 = 0$, they have $n_1 = -1.80$, $\beta_1^2 d^2 = 5.32$ and $\beta_2^2 d^2 = 0.86$.

This account follows [59, 61] closely. The final result is surprising, however, in view of the result proved earlier (which can also be deduced by transforming the formulae of *Dymnikov* et al.) that the lateral aberration at a line image of the (03)* or (30)* type, according to the line image in question, is never zero. This does not restrict the values of the virtual aberrations at a virtual line image, but is valid at any real line image, irrespective of the sequence of excitations and the geometry between object point and line image. For a comparable situation, see *Lenz* [122b]; cf. my "Real and virtual quadrupole aberrations" (to be published in *Optik*).

5.3. Disparities between the field distributions

a) Small differences. The effects of slight disparities between the z-dependences of $D(z)$ and $Q(z)$ are studied by *Shpak* and *Yavor* using the rectangular model in [189, 190], and summarized briefly in [239]. The distributions are presumed to differ in one of two ways: either their centres coincide but their effective lengths differ, or their effective lengths are the same while their centres are displaced.

(i) Centres coincident, lengths different. The common length is denoted by L and the amount by which the electrostatic length exceeds L is denoted by 2λ; the object distance, measured from the beginning of the lens, is denoted by u and the distance to the real line image (in the $x-z$ plane), measured from the end of the lens, by v. Then in image space, distance z from the end of the lens,

$$\frac{x}{x'_0} = (u+2\lambda)\cos\beta L + \frac{1}{\beta}\{1+(\beta_E^2-\beta^2)\,u\,\lambda\}\sin\beta L + \\ + z\left\{(1+2\beta_E^2\,u\,\lambda)\cos\beta L + \beta\left(-u+\frac{\beta_E^2-\beta^2}{\beta^2}\lambda\right)\sin\beta L\right\}$$

$$\frac{y}{y'_0} = (u+2\lambda)\,\mathrm{Ch}\,\beta L + \frac{1}{\beta}\{1+(\beta^2-\beta_E^2)\,u\,\lambda\}\,\mathrm{Sh}\,\beta L + \\ + z\left\{(1-2\beta_E^2\,u\,\lambda)\,\mathrm{Ch}\,\beta L + \beta\left(u+\frac{\beta^2-\beta_E^2}{\beta^2}\lambda\right)\mathrm{Sh}\,\beta L\right\}.$$

Setting $x=0$ yields $z=v$, and writing $\partial v/\partial\Phi = 0$, we obtain the achromatic condition for small values of λ:

$$\frac{\beta_E^2}{\beta_M^2} = \frac{1}{2}\left(1-\frac{\lambda}{L}f\right) \tag{5.14a}$$

with

$$f = f\left(\frac{u}{L},\beta L\right) \\ = \frac{2\left(\frac{u}{L}\right)^2 + \left\{\frac{1}{(\beta L)^2}-\left(\frac{u}{L}\right)^2\right\}\sin^2\beta L + \frac{1}{\beta L}\frac{u}{L}\sin 2\beta L}{\frac{1}{2}\left(\frac{u}{L}\right)^2 + \frac{1}{2(\beta L)^2} + \frac{1}{(\beta L)^2}\frac{u}{L}\sin^2\beta L + \frac{1}{4\beta L}\left\{\left(\frac{u}{L}\right)^2 - \frac{1}{(\beta L)^2}\right\}\sin 2\beta L} \tag{5.14b}$$

which simplifies to
$$f(\beta L) = \frac{3 + \cos 2\beta L}{1 + \frac{\sin 2\beta L}{2\beta L}} \tag{5.15}$$

at the focus. Since f is always positive (except when both $u = 0$ and $\sin\beta L = 0$), β_E^2 is now less than $\beta_M^2/2$ for positive values of λ and greater for negative values. The magnification is not affected by a small change in Φ if

$$\frac{\beta_E^2}{\beta_M^2} = \frac{1}{2}\left(1 - \frac{2\lambda}{L}\,\frac{2\frac{u}{L} - \frac{u}{L}\sin^2\beta L + \frac{1}{2\beta L}\sin 2\beta L}{\frac{u}{L} + \frac{1}{(\beta L)^2}\sin^2\beta L + \frac{1}{2\beta L}\frac{u}{L}\sin 2\beta L}\right).$$

This condition and condition (5.14) are satisfied simultaneously for a particular value of βL at each value of u; in [189], solutions for $u = 0$, L and $2L$ are given.

A line image can be made achromatic over its entire length; if for example, the line focus is considered, possible conditions for this are $\beta L = \pi/2$, $\beta L = 2.46$. (If the magnetic length exceeds the common length by λ, the sign of λ in these formulae is reversed.)

(ii) Displaced centres, equal lengths. The condition for achromatism is now

$$\frac{\beta_E^2}{\beta_M^2} = \frac{1}{2}\left(1 \pm \frac{\lambda}{L} f^*\right) \quad \begin{cases} +: \text{magnetic centre nearer object} \\ -: \text{electrostatic centre nearer object} \end{cases}$$

in which

$$f^* = f^*\left(\frac{u}{L}, \beta L\right) = -\frac{\left\{\frac{1}{(\beta L)^2} - \left(\frac{u}{L}\right)^2\right\}\sin^2\beta L + \frac{1}{\beta L}\frac{u}{L}\sin 2\beta L}{\frac{1}{2}\left(\frac{u}{L}\right)^2 + \frac{1}{2}\frac{1}{(\beta L)^2} + \frac{1}{(\beta L)^2}\frac{u}{L}\sin^2\beta L + \frac{1}{4\beta L}\left\{\left(\frac{u}{L}\right)^2 - \frac{1}{(\beta L)^2}\right\}\sin 2\beta L}$$

and λ now denotes the distance between the centres. The sign of f^* can change, and for

$$\frac{u}{L} = \frac{\cot\beta L/2}{\beta L}, \quad \frac{u}{L} = -\frac{\tan\beta L/2}{\beta L}$$

f^* vanishes.

These formulae are discussed further, and experimentally verified, in [190].

b) Large differences: separated lenses. The possibility of obtaining an achromatic system with separated electrostatic and magnetic quadrupoles is considered in detail by *Dymnikov* and *Yavor* [52] and mentioned in [239]. The general case of n quadrupole lenses, excitations $p_j = \pm\beta_j^2$ is considered and the chromatic aberration is shown to be of the form

$$v\frac{dg_n}{dv} = \sum_{j=1}^{n} m_j g_j\left[1 - \frac{1 + p_j g_j^2}{g_j}\left\{L_j + \frac{s_{j-1,j} - g_{j-1}}{1 + p_j(s_{j-1,j} - g_{j-1})^2}\right\}\right] \times$$
$$\times \prod_{i=j}^{n} \frac{1 + p_{i+1} g_{i+1}^2}{1 + p_{i+1}(s_{i,i+1} - g_i)^2}$$

in which v now represents velocity, $s_{j-1,j}$ is the drift space between the $(j-1)$-th and j-th lenses, and g_j denotes the distance from the end of the

j-th lens to the real or virtual point at which the emerging ray intersects the z-axis; m_j is a number equal to 1 if the j-th lens is electrostatic and equal to 1/2 if it is magnetic.

In the $x - z$ plane, a doublet can have zero or negative chromatic aberration only if the first (converging) lens is magnetic and the second is electrostatic. The opposite is true in the $y - z$ plane.

The paper concludes with a study of the conditions which must be satisfied in the design of an achromatic quadruplet.

We recall that *Meads'* analysis is designed to give chromatic aberration coefficients as well as geometrical and mechanical aberration coefficients [130]; the formulae are listed in § 3. The chromatic aberrations of the single quadrupole, the doublet and the triplet have been studied by *Regenstreif* [150] using the thin lens approximation for each lens.

6. Concluding remarks

One purpose of this review has been to indicate what properties of quadrupoles and quadrupole systems have been investigated, and by whom. In the past, there has been some duplication of results, especially among the theoreticians, and what is perhaps more serious, little comparison of freshly derived results and measurements with earlier attempts. I have taken some care, therefore, to indicate where future investigators can find work comparable with theirs, but as this is not intended to be a treatise on quadrupole lenses, I have not thought it proper to embark on elaborate comparisons and checks here. In discussing the work of *Meads*, only his methods of obtaining the aberration coefficients and the formulae have been described; this does not, however, do full justice to *Meads'* achievement, as he used the formulae as a basis for an IBM 7090 programme, designed to calculate the cardinal elements and aberrations of quadrupole systems, and subsequently improved and embellished for the IBM 7094. Reference [130] contains full details of this programme, which is described in [132a], and it will doubtless eventually "produce some sort of handbook of numbers for easy reference" (*Cohen*); it is to be hoped that this handbook will embrace both the rectangular and bell-shaped models.

It is to *Scherzer* that we owe the suggestion that quadrupoles and octopoles provide a likely means of overcoming spherical aberration; to conclude, therefore, let us recall the closing words of his 1947 article: "Wird trotzdem eine Voraussage erwartet, so möchte ich rein gefühlsmäßig den unrunden und den Hochfrequenzlinsen zutrauen, daß sie als erste eine Auflösung von einigen Ångström erreichen, die für das Sichtbarwerden schwerer Atome genügen dürfte. Welches Verfahren jedoch als erstes an die durch die Wärmebewegung gegebene Grenze des nützlichen Auflösungsvermögens heranführen wird, dürfte bei unseren heutigen Erfahrungen in keiner Weise zu prophezeien sein."

References

1. *Amboss, K.:* Electron optics: aberrations of air-cored magnetic lenses. Thesis, London 1959
2. *Anders, B., S. Hiekmann, H. Odrich,* u. *H. Trautmann:* Kernenergie 6, 63-65 (1963)
3. *Archard, G. D.:* Brit. J. Appl. Phys. **5**, 294—299 (1954)
4. — Proc. Conf. Electron Microscopy, London, 97—105 (1954)
5. — Proc. Phys. Soc. (London) B **68**, 156—164 (1955)
6. — Proc. Phys. Soc. (London) **72**, 135—137 (1958)
7. — *T. Mulvey,* and *D. P. R. Petrie:* Proc. Conf. Electron Microscopy, Delft, 51—54 (1960)
8. *Auberson, G.:* Design of beam transport systems. 1. Aperçu des méthodes de calcul. (CERN) MPS/Int. DL 61—36 (1961)

8a. *Audoin, C.:* Rev. Phys. Appl. **1**, 2—10 (1966)

8b. *Banford, A. P.:* The transport of charged particle beams. London: Spon 1966

9. *Basargin, Yu. G.,* i *N. I. Venikov:* Zhur. Tekh. Fiz. **34**, 1998—2002 (1964) = Soviet Phys. Tech. Phys. **9**, 1536—1539
10. *Bauer, H.-D.:* Naturwissenschaften **51**, 632 (1964); further details published in: Optik **23**, 596—609 (1965/6)
11. *Bernard, M.-Y.:* Compt. rend. **236**, 185—187 (1953)
12. — Compt. rend. **236**, 902—904 (1953)
13. — La «focalisation forte» dans les accélérateurs linéaires d'ions. Thèse, Paris 1953; published as Ann. phys. **9**, 633—682 (1954)
14. — Compt. rend. **240**, 1612—1616 (1955)
15. —, et *J. Hue:* Compt. rend. **243**, 1852—1854 (1956); see [210]
16. — — Compt. rend. **244**, 732—735 (1957)
17. *Blewett, J. P.:* On the design of quadrupole focusing systems. Brookhaven JPB-11 (1958)
18. — The focal properties of certain quadrupole lenses. Brookhaven JPB-13 (1959)

18a. *Bok, A. B., J. Kramer,* and *J. B. Le Poole:* Proc. Conf. Electron Microscopy, Prague, Appendix, 9 (1964)

18b. *Boussard, D.:* Trans. I.E.E.E. NS **12**, 648—651 (1965)

18c. —, and *A. Septier:* Trans. I.E.E.E. NS **12**, 652—655 (1965)

19. *Bruck, H.,* et *G. Gendreau:* Onde Elec. **35**, 1009—1021 (1955)
20. *Bullock, M. L.:* Am. J. Phys. **13**, 264—268 (1955)
21. *Burfoot, J. C.:* A contribution to the correction of aberrations in electrostatic lenses. Thesis, Cambridge 1952
22. — Proc. Phys. Soc. (London) B **66**, 775—792 (1953)
23. — Proc. Phys. Soc. (London) B **67**, 523—528 (1954)
24. — Proc. Conf. Electron Microscopy, London 105—109 (1954)
25. *Carathéodory, C.:* Geometrische Optik. Ergeb. Math. Grenzgeb. **4**, No 5 (1937)
26. *Carlile, R. N.:* The quadrupole magnet as a focusing device. Stanford HEPL-33 (1957)
27. *Chako, N.:* Proc. McGill University Symp. on Microwave Optics. Vol. I. 67—82 (1953, published 1959)
28. — Trans. Chalmers Univ. Technol. Gothenburg No. 191, 50 pp. (1957)
29. *Chamberlain, O.:* Ann. Rev. Nuclear Sci. **10**, 161—192 (1960)
30. *Christofilos, N.:* Focusing system for ions and electrons and application in magnetic resonance particle accelerators. Unpublished, 1950 (see [35])
31. *Cohen, D.:* Review and discussion of quadrupole magnet aberrations. Unpublished, 1965
32. *Cork, B.,* and *E. Zajec:* Phys. Rev. **92**, 853 (1953)
33. *Cotte, M.:* Recherches sur l'optique électronique. Thèse, Paris 1938; published in Ann. phys. (11) **10**, 333—405 (1938)
34. *Courant, E. D., M. S. Livingston,* and *H. S. Snyder:* Phys. Rev. **88**, 1190—1197 (1952)
35. — — —, and *J. P. Blewett:* Phys. Rev. **91**, 202—203 (1953)
36. *Crewe, A. V.:* Some effects of quadrupole misalignment in a microscope. Unpublished, 1965

36a. — Proc. Conf. Electron Microscopy, Kyoto 627 (1966)
37. *Dayton, I. E., E. C. Shoemaker*, and *R. F. Mozley:* Rev. Sci. Instr. **25**, 485—489 (1954)
38. *Deltrap, J. H. M.*, and *V. E. Cosslett:* Proc. Conf. Electron Microscopy, Philadelphia, KK-8 (1962)
39. — Correction of spherical aberration of electron lenses. Thesis, Cambridge 1964
40. — Proc. Conf. Electron microscopy, Prague, A 45—46 (1964)
41. *Dhuicq, D.*, et *A. Septier:* Compt. rend. **249**, 2031—2033 (1959)
42. — Compt. rend. **251**, 1989—1991 (1960)
43. — Recherche de systèmes de lentilles quadrupolaires stigmatiques au 1er ordre. Thèse 3e cycle, Paris 1961
43a. —, et *A. Septier:* Etude expérimentale et correction des aberrations d'ouverture d'un triplet électrostatique à focalisation forte stigmatique au premier ordre. Colloque annuel, SFME, Strasbourg 1964
43b. — — Sur la possibilité de réaliser un objectif formé de quatre lentilles quadrupolaires. Colloque annuel, SFME, Marseille 1965
43c. — — Compt. Rend. **263 B** 280—283 (1966)
43d. — — Compt. Rend. **263 B** 364—367 (1966)
44. *Draper, J.:* Rev. Sci. Instr. **34**, 679—684 (1963)
45. *Dušek, H.:* Über die elektronenoptische Abbildung durch Orthogonalsysteme unter Verwendung eines neuen Modellfeldes mit streng angebbaren Paraxialbahnen. Dissertation, Vienna 1958
46. — Optik **16**, 419—445 (1959)
47. *Dutov, G. G., A. M. Solov'ev*, i *S. A. Toporkov:* Izvest. Akad. Nauk S.S.S.R. Ser. Fiz. **27**, 1154-1157 (1963) = Bull. Acad. Sci. U.S.S.R. Phys. Ser. **27**, 1135-1137
48. *Dymnikov, A. D., T. Ya. Fishkova*, i *S. Ya. Yavor:* Izvest. Akad. Nauk S.S.S.R. Ser. Fiz. **27**, 1131—1134 (1963) = Bull. Acad. Sci. U.S.S.R. Phys. Ser. **27**, 1112—1115
49. —, *L. P. Ovsyannikova*, i *S. Ya. Yavor:* Zhur. Tekh. Fiz. **33**, 393—397 (1963) = Soviet Phys. Tech. Phys. **8**, 293—296
50. —, i *S. Ya. Yavor:* Zhur. Tekh. Fiz. **33**, 851—858 (1963) = Soviet Phys. Tech. Phys. **8**, 639—643
51. — *T. Ya. Fishkova* i *S. Ya. Yavor:* Zhur. Tekh. Fiz. **34**, 1711—1714 (1964) = Soviet Phys. Tech. Phys. **9**, 1322—1334
52. —, i *S. Ya. Yavor:* Zhur. Tekh. Fiz. **34**, 2008—2014 (1964) = Soviet Phys. Tech. Phys. **9**, 1544—1548
53. —, *T. Ya. Fishkova* i *S. Ya. Yavor:* Doklady Akad. Nauk S.S.S.R. **154**, 1321—1324 (1964) = Soviet Phys. Doklady **9**, 182—185
54. — — — Proc. Conf. Electron Microscopy, Prague, A 43—44 (1964)
55. — — — Radiotekh. i Elektron. **9**, 1828—1831 (1964) = Radio Eng. and Electronics (U.S.S.R.) **9**, 1515—1518
56. — — — Doklady Akad. Nauk S.S.S.R. **162**, 1265—1268 (1965) = Soviet Phys. Doklady **10**, 547—550
57. — — — Zhur. Tekh. Fiz. **35**, 431—440 (1965) = Soviet Phys. Tech. Phys. **10**, 340—346
58. — — — Zhur. Tekh. Fiz. **35**, 759—761 (1965) = Soviet Phys. Tech. Phys. **10**, 592—594
59. — — — Zhur. Tekh. Fiz. **35**, 1068—1076 (1965) = Soviet Phys. Tech. Phys. **10**, 822—827
60. — — — Phys. Letters **15**, 132—134 (1965)
60a. — — — Nuclear Instr. & Methods **37**, 268—275 (1965)
60b. —, and *S. Ya. Yavor:* Nuclear Instr. & Methods **40**, 161—162 (1966)
60c. —, *T. Ya. Fishkova* i *S. Ya. Yavor:* Izvest. Akad. Nauk S.S.S.R. Ser. Fiz. **30**, 739—741 (1966) = Bull. Acad. Sci. U.S.S.R. Phys. Ser. **30**
60d. — — *L. P. Ovsyannikova* i *S. Ya. Yavor:* Nucl. Instr. & Methods, **42**, 293 to 296 (1966)
61. *Elmore, W. C.*, and *M. W. Garrett:* Rev. Sci. Instr. **25**, 480—485 (1954)
61a. *Emmerson, J. McL.*, and *N. Middlemas:* Nucl. Instr. & Methods, **24**, 93-102 (1963)
62. *Enge, H. A.:* Rev. Sci. Instr. **30**, 248—251 (1959)
63. — Rev. Sci. Instr. **32**, 662—665 (1961)

64. *Fert, C.*, et *R. Saporte:* Compt. rend. **243**, 1107—1110 (1956)
64a. *Fishkova, T. Ya., L. P. Ovsyannikova*, and *S. Ya. Yavor:* Proc. Conf. Electron Microscopy, Kyoto 209—210 (1966)
65. *Focke, J.:* Jenaer Jahrb. 69—78 (1952)
66. — Progr. Opt. **4**, 1—36 (1965)
67. *Geibel, J. A.*, and *N. M. King:* Given transformation matrices related to a quadrupole and drift lengths. (CERN) MPS/EP/63-2 (1963)
68. *Giese, C. F.:* Rev. Sci. Instr. **30**, 260—261 (1959)
69. *Glaser, W.:* Z. Physik **120**, 1—15 (1943)
70. —, and *O. Bergmann:* Z. angew. Math. u. Phys. **1**, 363—379 (1950)
71. — Grundlagen der Elektronenoptik. Vienna: Springer 1952 (*q.v.* for a full account of *Glaser's* earlier work)
72. —, u. *P. Schiske:* Optik **11**, 422—443, and 445—467 (1954)
73. — Investigations of four-pole lenses; an electron microscope corrected with respect to spherical aberration. Farrand Optical Company Internal Reports (1955). Many details are to be found in "Electron Lens", U. S. Patent No. 2, 919, 381 (Application: July 25, 1956; patented: Dec. 29, 1959)
74. — Elektronen- und Ionenoptik. Handbuch der Physik **33**, 123—395. Berlin: Springer 1956, quadrupoles are dealt with in § 37 "Fokussierung in Feldern ohne Rotationssymmetrie", pp. 257—272
75. *Gregory, B. C.*, and *K. F. Sander:* J. Elec. Control **13**, 123—136 (1962)
76. *Grinberg, G. A.* (= *Grünberg*): Doklady Akad. Nauk S.S.S.R. **37**, 172—178 and 261—268 (1942), and **38**, 78—81 (1943) = Zhur. Tekh. Fiz. **13**, 361—388 (1943)
77. — Selected problems in the mathematical theory of electric and magnetic phenomena (in Russian). Izd. Akad. Nauk S.S.S.R. Moscow/Leningrad, 1948
78. — Optika i Spektroskopiya **3**, 673 (1957) and Zhur. Tekh. Fiz. **27**, 2425—2431 (1958) = Soviet Phys. Tech. Phys. **2**, 2259—2265
79. *Grivet, P.*, et *M.-Y. Bernard:* Aberrations des lentilles à forte convergence. Influence du champ longitudinal B_z. (CERN) PS/MB-6 (1955)
80. — *A. Septier* et *J. Hue:* Etude de lentilles magnétiques à focalisation forte. CERN Symposium, I, 192—199 (1956)
81. — — Les lentilles quadrupolaires magnétiques. CERN 58—25 (1958)
82. — — Nucl. Instr. & Methods **6**, 126—156, and 243—275 (1960)
83. — Electron optics. Oxford: Pergamon 1965; Strong-focusing lenses are dealt with in Chapter 10, pp. 233—258
85. *Hawkes, P. W.:* Proc. Conf. Electron Microscopy, Philadelphia, KK-7 (1962)
86. —, and *V. E. Cosslett:* Brit. J. Appl. Phys. **13**, 272—279 (1962)
87. — The aberrations of electron optical systems in the absence of rotational symmetry. Thesis, Cambridge 1963
88. — Proc. Conf. Electron Microscopy, Prague, A 5—6 (1964)
89. — Phil. Trans. Roy. Soc. London, Ser. A **257**, 479—522, and 523—552 (1965)
90. — Optik **22**, 349—368 (1965) and **24**, 94 (1966/7)
91. — Optik **22**, 543—551 (1965)
92. — Optik **23**, 145—168 (1965/66)
93. — Optik **23**, 244—250 (1965/66)
94. — Optik **24**, 60—78 and 95—107 (1966/7)
95. — Optik **24**, 252—262 (1966/7)
96. — "Lens aberrations". In: The focusing of charged particles (Ed. *Septier*). New York: Academic Press 1967
97. — Proc. Conf. Electron Microscopy, Kyoto, 205—206 (1966)
98. *Herzberger, M.:* Modern geometrical optics. New York: Interscience 1958
99. *Hubbard, E. L.*, and *E. L. Kelly:* Rev. Sci. Instr. **25**, 737—739 (1954)
100. *Hultschig, H.:* Zur Entwicklung der Quadrupollinsen I, II. DESY-Notiz A 2, 73 (1960), and DESY-Notiz A 2, 76 (1961)
101. *Iliescu, C. C.:* Nucl. Instr. & Methods **21**, 136—144, and 145—154 (1963); Stud. Cerc. Fiz. **16**, 1131—1206 (1964); Trans. I.E.E.E. NS **12**, 387—391 (1965)
102. *Judd, D. L.:* Ann. Rev. Nuclear Sci. **8**, 181—216 (1958)
103. *Junior, P.:* Z. angew. Phys. **18**, 152—157 (1964)
103a. *Kartashev, V. P.*, i *V. I. Kotov:* Zhur. Tekh. Fiz. **36**, 1569—1574 (1966) = Soviet Phys. Tech. Phys. **11**

104. *Kas'yankov, P. P.:* The theory of electromagnetic systems with a curved axis (in Russian). Izd. LGU, Leningrad 1956
105. — Methods of calculation for electron optical systems (in Russian). Thesis, Leningrad Electrotechnical Institute 1957
106. — Optika i Spektroskiya **3**, 169—179 (1957)
107. — Zhur. Tekh. Fiz. **28**, 915—918 (1958) = Soviet Phys. Tech. Phys. **3**, 854—857
108. *Kel'man, V. M.*, i *S. Ya. Yavor:* Zhur. Tekh. Fiz. **31**, 1439—1442 (1961) = Soviet Phys. Tech. Phys. **6**, 1052—1054
109. — — Zhur. Tekh. Fiz. **33**, 368—370 (1963) = Soviet Phys. Tech. Phys. **8**, 271—272
110. — — *A. D. Dymnikov*, i *L. P. Ovsyannikova:* Izvest. Akad. Nauk S.S.S.R. Ser. Fiz. **27**, 1135—1138 (1963) = Bull. Acad. Sci. U.S.S.R. Phys. Ser. **27**, 1116—1119
111. — — Electron optics (in Russian). Izd. Akad. Nauk S.S.S.R. Moscow/Leningrad, 2nd ed., 1963; for quadrupoles, see § 2 of Chapter VIII, pp. 234—255
112. *King, N. M.:* Some focusing properties of quadrupole doublets. (CERN) MPS/EP/22 (1962)
113. — Doublet configuration representing given transformation matrices. (CERN) MPS/Int/EP 63—4 (1963)
114. — Focusing with quadrupole triplets: thin lens approximation. (CERN) MPS/EP/63—6 (1963)
115. — Progr. Nuclear Phys. **9**, 71—116 (1964)
116. *Klemperer, O.:* Improvements in electron lens systems. Brit. Pat. 573,901 (Applied 1942, accepted 1945) and Improvements in electron lenses. Brit. Pat. 574,056 (Applied 1942, accepted 1945)
117. — Electron optics. Cambridge: University Press 1953; concerning [116] see pp. 292—300
118. *Knowles, H. B.:* Nuclear Instr. & Methods **25**, 29—39 (1963)
119. *Koltay, E.:* Fizikai Szemle **9**, 182—187 (1959)
120. —, és *L. Fejer:* Atomki Közlemények **4**, 177—182 (1962)
121. —, és *Gy. Szabó:* Atomki Közlemények **6**, 105—129 (1964)
122. — — Nuclear Instr. & Methods **35**, 88—92 (1965)
122a. *Lapostolle, P.*, et *F. Fer:* Compt. rend. **256**, 5294—5297 (1963)
122b. *Lenz, F.:* Proc. Conf. Electron Microscopy, Stockholm, 48—51 (1956) and Optik **14**, 74—82 (1957)
122c. *Le Poole, J. B.:* Proc. Conf. Electron Microscopy, Prague, Appendix, 8 (1964); Proc. Conf. Electron Microscopy in Far East and Oceania, Calcutta 28—29 (1965)
123. *Lévy, P.:* J. phys. radium. **17**, 60a—61a (1956)
124. *Loebenstein, H. M.*, and *A. Sprinzak:* Nuclear Instr. & Methods **25**, 162—170 (1963); see [60b]
124a. *Lu, C.-S.*, and *H. E. Carr:* Rev. Sci. Instr. **33**, 823—824 (1962)
125. *Lublow, D.:* Zur Berechnung des Polschuhprofils bei magnetischen Quadrupollinsen. DESY-Notiz A 2, 78 (1961)
126. *Luckey, D.:* "Beam Optics". In: Techniques of High Energy Physics (Ed. *Ritson*). New York: Interscience 1961, 403—463
127. *Luneburg, R. K.:* Mathematical theory of optics (1944); published California U. P., 1965
128. *Lynch, P. J.*, and *D. J. Zaffarano:* Tests and analysis of magnetic quadrupole lenses. ISC-927 (1957)
128a. *Mal'tsev, A. P.*, i *V. A. Teplyakov:* Pribory i Tekh. Eksp. No. **4**, 29—31 (1965) = Instr. exp. Tech. 763—765
129. *Markovich, M. G.*, i *I. I. Tsukkerman:* Zhur. Tekh. Fiz. **30**, 1362—1368 (1960) = Soviet Phys. Tech. Phys. **5**, 1292—1298
130. *Meads, P. F.:* The theory of aberrations of quadrupole focusing arrays. Thesis, California, 1963; distributed as UCRL-10807
131. — The effects of misalignment of quadrupole magnets in an arbitrary beam system. MURA Technical Note, TN-513 (1964); to be published in Rev. Sci. Instr.
132. — Proc. AMU-ANL High voltage Electron Microscope Meeting, Argonne 82—92 (1964)

132a. — Nuclear Instr. & Methods **40**, 166—168 (1966)
133. *Melkich, A.:* Ausgezeichnete astigmatische Systeme der Elektronenoptik. Dissertation, Berlin, 1944; published in: Sitzber. Akad. Wiss. Wien, Abt. II A, **155**, 393—438, and 440—471 (1947)
134. *Meyer, W. E.:* Das Auflösungsvermögen sphärisch korrigierter elektrostatischer Elektronenmikroskope. Dissertation, Darmstadt 1958; parts published as [135] and [136]
135. — Optik **18**, 69—91 (1961)
136. — Optik **18**, 101—114 (1961)
137. *Miller, V. V.:* Pribory i Tekh. Eksp. No. 4, 23—25 (1964) = Instr. exp. Tech. 743—744
138. — Pribory i Tekh. Eksp. No. 6, 3—23 (1964) = Instr. exp. Tech. 1141—1161
139. *Möllenstedt, G.:* Proc. Conf. Electron Microscopy, London 694—698 (1954)
140. — Optik **13**, 209—215 (1956)
140a. *Möller, P-A., D. Dhuicq* et *A. Septier:* Compt. Rend. **263**, 534—537 (1966)
140b. *Moses, R. W.:* Rev. Sci. Instr. **37**, 1370—1372 (1966)
141. *Orr, D.:* Electron model study of particle motions in electrostatic quadrupoles. Thesis, London 1961
142. — Nuclear Instr. & Methods **24**, 377—380 (1963)
143. *Ovsyannikova, L. P.,* i *S. Ya. Yavor:* Zhur. Tekh. Fiz. **35**, 940—946 (1965) = Soviet Phys. Tech. Phys. **10**, 723—726
144. *Penner, S.:* Rev. sci. Instr. **32**, 150—160 (1961)
145. *Picht, J.:* Einführung in die Theorie der Elektronenoptik. Leipzig: Barth 1939 (1st ed.), and 1963 (3rd ed.); for quadrupoles, see § 25, pp. 222—229 and § 35, pp. 271—275
146. *Pilat, I. M., E. E. Reznik* i *A. M. Strashkevich:* Zhur. Tekh. Fiz. **21**, 1149—1152 (1951)
147. *Plotnikov, V. K.:* Pribory i Tekh. Eksp. No. 2, 29—33 (1962) = Instr. exp. Tech., 246—251
148. *Rabin, B. M., A. M. Strashkevich* i *L. S. Khin:* Zhur Tekh. Fiz. **21**, 438—447 (1951)
149. *Recknagel, A.:* Wiss. Z. TH Dresden **8**, 933—939 (1959)
150. *Regenstreif, E.:* Chromatic aberration in quadrupole multiplets. CERN 64—72 (1963)
151. — Quadrupole lenses. In: The focusing of charged particles (Ed. *Septier*). New York: Academic Press 1967
152. *Reisman, E.:* The imaging properties of crossed magnetic quadripoles. Thesis, Cornell 1957
153. —, and *B. M. Siegel:* Bull. Am. Phys. Soc. (2) **2**, 174 (1957)
153a. *Rose, H.:* Optik **24**, 36—59 and 108—121 (1966/67)
154. *Rosenblatt, J.:* Nuclear Instr. & Methods **5**, 152—155 (1959)
154a. *Sacerdoti, G.,* e *F. Uccelli:* Suppl. Nuovo Cim. (10) **26**, 238—250 (1962)
155. *Scherzer, O.:* Z. Physik **101**, 593—603 (1936)
156. — Optik **2**, 114—132 (1947)
157. — Optik **5**, 497—498 (1949)
158. — Proc. Conf. Electron Microscopy, Paris 191—195 (1950)
159. — Optik **22**, 314—318 (1965)
160. *Schiske, P.:* Nature (London) **171**, 443—444 (1953), and Acta Phys. Austriaca **6**, 221 (1952/53)
161. *Schleich, F.:* Proc. Conf. Electron Microscopy, Delft, **1**, 48—50 (1960)
162. —, u. *D. Hoffmeister:* Über die Feldverteilung in Vierpol-Linsen auf Grund von Präzisionsmessungen. Tagung für Elektronenmikroskopie, Aachen 1965
163. —, u. *H. Koops:* Zur Berechnung der Kaustikfiguren von 8-Pol-Linsen. Tagung für Elektronenmikroskopie, Aachen 1965
164. *Schneider, H.:* Nuclear Instr. & Methods **1**, 268—273 (1957)
165. *Schüler, G.:* Unpublished Dresden Diplomarbeit, see [149]
166. *Seeliger, R.:* Optik **4**, 258—262 (1948/49); **5**, 490—496 (1949); **8**, 311—317 (1951); **10**, 29—41 (1953)
167. *Seman, O. I.:* Doklady Akad. Nauk S.S.S.R. **81**, 775—778 (1951)
168. — Doklady Akad. Nauk S.S.S.R. **96**, 1151—1154 (1954)
169. — Trudy Inst. Fiz. i Astron. Akad. Nauk Eston. S.S.R. No. 2, 3—49 (1955)

170. — Uchenye Zapiski Rostovsk. Gosudarst. Univ. Ser. Fiz. **68**, 63—75, and 77—90 (1958)
171. *Septier, A.:* Compt. rend. **243**, 132—135 (1956)
172. — Compt. rend. **243**, 1026—1029 (1956)
173. — Compt. rend. **243**, 1297—1300 (1956)
174. — Compt. rend. **245**, 1406—1409 (1957)
175. — Compt. rend. **245**, 1905—1908 (1957)
176. — Compt. rend. **245**, 2036—2039 (1957)
177. — Compt. rend. **246**, 1835—1838 (1958)
178. — Compt. rend. **246**, 1983—1985 (1958)
179. — Proc. Conf. Electron Microscopy, Berlin, Vol. I, 57—61 (1958)
180. —, et *J. van Acker:* Compt. rend. **251**, 346—348, and 1750—1752 (1960)
181. — — Proc. Conf. Electron Microscopy, Delft, Vol. I, 44—47 (1960)
182. — Advances in Electronics and Electron Phys. **14**, 85—205 (1961)
183. — Compt. rend. **252**, 2851—2853 (1961)
184. —, and *J. van Acker:* Nuclear Instr. & Methods **13**, 335—355 (1961)
185. — Proc. Conf. Electron Microscopy, Philadelphia KK-10 (1962)
186. — Compt. rend. **256**, 2325—2328 (1963), and J. Microscopie **2**, 17 (1963)
187. — Advances Opt. and Electron Microscopy **1**, 204—274 (1966)
188. *Shpak, E. V.*, i *S. Ya. Yavor:* Zhur. Tekh. Fiz. **34**, 1037—1039 (1964) = Soviet Phys. Tech. Phys. **9**, 803—805
189. — — Zhur. Tekh. Fiz. **34**, 2003—2007 (1964) = Soviet Phys. Tech. Phys. **9**, 1540—1543
190. — — Zhur. Tekh. Fiz. **35**, 947–950 (1965) = Soviet Phys. Tech. Phys. **10**, 947–950
191. *Shukeilo, I. A.:* Zhur. Tekh. Fiz. **29**, 1225—1227 (1959) = Soviet Phys. Tech. Phys. **4**, 1123—1125
192. *Shull, F. G., C. E. McFarland*, and *M. M. Bretscher:* Rev. Sci. Instr. **25**, 364—367 (1954)
193. *Siegel, B. M.*, and *E. Reisman:* J. Appl. Phys. **25**, 1453 (1954)
194. *Sonoda, M., A. Katase, M. Seki*, and *T. Akiyoshi:* Nuclear Instr. & Methods **12**, 349—352 (1961)
195. *Stedman, E. C.:* Design of beam transport systems. 2. Practical beam design and setting up, including properties of selected beams. (CERN) MPS/Int. DL 62—23 (1962)
196. *Steffen, K. G.:* A quadrupole magnet with non-circular aperture and linearized end fringing field. DESY-Notiz A 2.81 (1961)
197. — High energy beam optics. New York: Interscience 1965
198. *Sternheimer, R. M.:* Beam transport systems. In: Methods of Experimental Physics. **5** B, 691—747. New York: Academic Press 1963
199. *Strashkevich, A. M.*, i *I. M. Pilat:* Izvest. Akad. Nauk S.S.S.R. Ser. Fiz. **15**, 448—466 (1951)
200. — — Chernovits. Gos. Univ.: Uchen. Zap. (Ser. Fiz.-mat. Nauk) **4**, 113—122 (1952)
201. — Zhur. Tekh. Fiz. **22**, 487—497 (1952)
202. — Zhur. Tekh. Fiz. **22**, 1848—1856 (1952)
203. — Zhur. Tekh. Fiz. **24**, 274—286 (1954)
204. —, i *N. G. Gluzman:* Zhur. Tekh. Fiz. **24**, 2271—2284 (1954)
205. — Dopovidi Akad. Nauk Ukr. R.S.R. 929—932 (1958)
206. — The electron optics of electrostatic fields not possessing axial symmetry (in Russian). Moscow: Fizmatgiz 1959
207. — Izvest. Akad. Nauk S.S.S.R. Ser. Fiz. **23**, 706—710 (1959) = Bull. Acad. Sci. U.S.S.R. Phys. Ser. **23**, 701—705
208. — Radiotekh. i Elektron. **5**, 1997—2003 (1960) = Radio Eng. and Electronics (U.S.S.R.) **5**, No. 12, 181—191
209. — Zhur. Tekh. Fiz. **30**, 1199-1206 (1960) = Soviet Phys. Tech. Phys. **5**, 1136-1142
210. — Radiotekh. i Elektron. **6**, 1562—1565 (1961) = Radio Eng. and Electronics (U.S.S.R.) **6**, 1392—1396
211. — Radiotekh. i Elektron. **6**, 1725—1728 (1961) = Radio Eng. and Electronics (U. S.S.R.) **6**, 1538—1541
212. — Zhur. Tekh. Fiz. **33**, 512—517 (1963) = Soviet Phys. Tech. Phys. **8**, 380—383

213. — Zhur. Tekh. Fiz. **34**, 1401—1408 (1964) = Soviet Phys. Tech. Phys. **9**, 1082—1086
213a. — Radiotekh. i Elektron. **10**, 1512—1517 (1965) = Radio Eng. and Electronics **10**, 1300—1305
214. *Sturrock, P. A.:* Proc. Roy. Soc. (London) A **210**, 269—289 (1951)
215. — Hamiltonian Electron Optics. Thesis, Cambridge 1951
216. — Phil. Trans. Roy. Soc. London. Ser. A **245**, 155—187 (1952)
217. — Static and dynamic electron optics. Cambridge: University Press 1955
218. *Tanguy, P.:* Compt. rend. **261**, 1811—1813 (1965)
219. — Compt. rend. **261**, 1945—1947 (1965)
220. *Tawara, H.*, and *S. Suganomata:* Hitachi Hyoron **47**, 677—681 (1965)
220a. *Teplov, I. B.*, i *L. N. Fateeva:* Pribory i Tekh. Eksp. No. 6, 45—51 (1965) = Instr. exp. Tech. 1345—1351
221. *Thomson, M. G. R.:* Proc. Conf. Electron Microscopy, Kyoto 206—207 (1966)
222. *Timm, U.*, u. *H. Schneemann:* Auslegung und Konstruktion von Quadrupoltripeln für die Injektion. DESY-Notiz A 2, 47 (1959)
223. *Tsukkerman, I. I.:* Zhur. Tekh. Fiz. **24**, 258—273 (1954)
224. — Zhur. Tekh. Fiz. **24**, 2261—2263 (1954)
225. — Electron optics in television. Oxford: Pergamon 1961 (a translation of the Russian text, published in 1958)
226. — Zhur. Tekh. Fiz. **28**, 1809—1812 (1958) = Soviet Phys. Tech. Phys. **3**, 1668—1670
227. — Zhur. Tekh. Fiz. **33**, 505—511 (1963) = Soviet Phys. Tech. Phys. **8**, 375—379
227a. *Uccelli, F.:* Calcolo delle matrici di trasfert di un quadrupolo col modello trapezoidale. LNF Report (1962); see [154a]
228. *Vandakurov, Yu. V.:* Zhur. Tekh. Fiz. **27**, 1850—1862 (1957) = Soviet Phys. Tech. Phys. **2**, 1719—1733
228a. *Vash, A. M.:* The design and use of alternating gradient lenses. M. S. Thesis, Massachusetts Institute of Technology 1953
229. *Venikov, N. I.*, i *E. S. Mironov:* Zhur. Tekh. Fiz. **34**, 530—535 (1964) = Soviet Phys. Tech. Phys. **9**, 413—416
230. *Vivargent, M.:* Focalisation et analyse du faisceau du cyclotron du Collège de France. Utilisation du faisceau analysé de particules α à l'excitation coulombienne de noyaux lourds. Thèse, Paris, 1958; published in Ann. phys. (13) **4**, 1047—1109 (1959)
231. *Weidemann, W.:* Nuclear Instr. & Methods **9**, 347—353 (1960)
232. *Whitmer, R. F.:* An investigation of non-rotationally symmetrical electrostatic electron optical lenses. Thesis, Polytechnic Institute of Brooklyn 1956
233. — J. Appl. Phys. **27**, 808—815 (1956)
234. *Yagi, K.:* A simple method for evaluations of the aberrations due to quadrupole-magnet systems. Tokyo INSJ-78 (1964); cf. [60c]
235. — Nuclear Instr. & Methods **31**, 173—188 (1964), and **34**, 146—154 (1965); cf. [60c]
236. *Yavor, S. Ya.:* Proc. Symposium Electron and Vacuum Physics, Budapest, 125—137 (1962)
237. —, i *A. D. Dymnikov:* Doklady Akad. Nauk S.S.S.R. **154**, 821—823 (1964) = Soviet Phys. Doklady **9**, 177—178
238. — — and *L. P. Ovsyannikova:* Nuclear Instr. & Methods **26**, 13—17 (1964)
239. — — *T. Ya. Fishkova*, and *E. V. Shpak:* Proc. Conf. Electron Microscopy, Prague, A 35—36 (1964)
240. — — i *L. P. Ovsyannikova:* Zhur. Tekh. Fiz. **34**, 99—104 (1964) = Soviet Phys. Tech. Phys. **9**, 76—80

Strashkevich, A. M.: The electron optics of electrostatic systems (in Russian). Moscow/Leningrad: Energiya 1966 (especially Chapters 8—11)

Dr. *P. W. Hawkes*, Research Fellow of Peterhouse,
Electron Microscope Section, Cavendish Laboratory, Free School Lane,
Cambridge, England